The Shattered Cross

The Rise, Fall and Undying Legacy of the Knights Templars

Allen Schery

Brooklyn Bridge Books

Copyright ©2025 by Allen Schery

All rights reserved.

ISBN: 978-1-968950-02-6

No portion of this book may be reproduced in any form without written permission from the publisher or author, except as permitted by U.S. copyright law.

Dedication

I dedicate this work to my initial mentors in History: Romaine Francois Poirot and Alice D'Addario at Walt Whitman High School and Regis Courtemanche and Benjamin Ruekberg at C.W. Post College. Although I moved into Anthropology and Archeology this scholarly work in history shows their undying influence and I celebrate their essence and eternally thank them.

Contents

1. Origins of the Templars — 1
2. Rise to Power & Military Influence — 7
3. Spires of Sacrifice: The Templars' Legacy in Jerusalem — 13
4. Templar Architecture and Fortifications — 19
5. Wealth, Banking, and Political Power — 27
6. Secret Rites, Symbols & Allegations — 32
7. King Philip IV & the Fall of the Templars — 39
8. The Escape & Survival Theories — 44
9. The Hidden Templar Treasure — 49
10. The Fleet of La Rochelle & Possible Exile — 55
11. The Oak Island Mystery — 59
12. The Mystery of Rosslyn Chapel — 74
13. The Hospitallers & The Fate of Templar Assets — 81
14. Templar Influence on Secret Societies & Freemasonry — 91
15. The Vatican Archives & Missing Templar Documents — 99
16. The Templar Curse & Unusual Deaths — 109

17. Lost Templar Ships & Theories of an Underwater Hoard	115
18. Were the Templars the First Global Intelligence Network?	122
19. The Templar Legacy	127
20. Bridging Innovation, Secrecy, and Modern Myth	133
About the Author	137
Bibliograpy	139
Index	152
Endnotes	164

Chapter One

Origins of the Templars

In the stormy tapestry of medieval Europe, the Templar Order emerged as a transcendent beacon of martial valor and spiritual devotion—a confluence of noble ambition and divine inspiration woven from religious fervor and calculated political strategy. Their inception was not the product of a singular impulse but rather the culmination of an era where sacred ideals merged with the imperatives of feudal duty. Amidst a continent defined by shifting alliances and ceaseless conflict, the early Templars gathered—a diverse brotherhood born of French, Italian, Spanish, and English traditions, where chivalric honor and political acumen were imbued in every knight.

Many of the founding members, such as Hugh de Payen's and Godfrey de Saint-Omer, hailed from aristocratic families steeped in devout religious tradition and the responsibilities of leadership on the battlefield and in the courts. Raised in an environment where piety was as important as martial training, these men envisioned a grand design: an Order to protect pilgrims and secure a sacred destiny. Their noble lineage, coupled with the eventual endorsement from influential figures like Bernard of Clairvaux and formal recognition at the Council of Troyes, laid the groundwork for an institution that transcended personal gain—a brotherhood driven by a commitment to both heavenly ideals and the pragmatic necessities of power. Significantly, their formal title, *Pauperes Commilitones Christi Templique Salomonici* (The Poor Fellow-Soldiers of Christ and the Temple of Solomon), underscored their initial vows of poverty and spiritual dedication, resonating deeply with the Christian faithful across Europe.

The choice of Jerusalem as the Order's epicenter was a revelation and a calculated strategy. On the Temple Mount, believed to be the site of Solomon's Temple, the early Knights were notably quartered, leading to theories of their search for ancient relics or hidden knowledge within its sacred confines. Jerusalem—steeped in millennia of sacred history, prophetic legacy, and divine promise—stood as the radiant heart of Christian redemption. For medieval believers, the city was far more than a conquest; it was the ultimate pilgrimage destination—a sacred nexus where the Passion and Resurrection of Christ converged with an enduring holiness tradition. Its ancient walls and hallowed landmarks beckoned the faithful toward an encounter with the divine while offering a potent geopolitical prize amid the turbulent currents of territorial ambition. Defending Jerusalem thus became emblematic of the Templars' dual quest: to preserve the sanctity of holy history and to initially secure a formidable stronghold for Christendom by safeguarding the perilous routes pilgrims took.

As the Order expanded its ranks, knights from every corner of Europe—French, Italian, Spanish, and English—lent their unique strengths to this growing brotherhood. The French, steeped in gallantry and fervent piety, often led the charge, while Italian recruits, renowned for their administrative brilliance and mercantile acumen, provided the organizational backbone. Spanish warriors, fueled by the rejuvenated spirit of the Reconquista, infused the Order with a fiery zeal, and English knights contributed a measured pragmatism that harmonized raw ambition with strategic restraint. This diverse cultural amalgam bolstered the Templar military might and forged a unity that transcended local rivalries and instilled the Order with a collective identity.

The Templars' distinctive attire was central to this identity—a striking visual manifesto of their sacred ideals. Draped in pristine white mantles, the knights intentionally renounced worldly opulence in favor of simplicity that evoked purity, self-denial, and transcendence. The white mantle, a canvas, was boldly emblazoned with a vivid red cross—a symbol piercing the fabric like a clarion call. This emblem, evoking the blood of Christ and the ultimate sacrifice, was more than decorative: it was a rallying sign, uniting knights from disparate lands under a cause as divine as it was militarily vital.

The theological justification for this new form of knighthood was powerfully articulated by Bernard of Clairvaux, who, in his influential treatise *De Laude Novae Militae* (In Praise of the New Knighthood), argued that killing in defense of the faith was not a sin but a righteous act that sanctified the warrior, thus providing a crucial spiritual foundation for the Order's unique mission. Their formal establishment was solidified at the Council of Troyes in 1129, where the foundational Rule of the Templars was officially promulgated, laying down the strict guidelines for their monastic and military life. Furthermore, their growing influence was cemented by crucial Papal Bulls, such as *Omne Datum Optimum* (1139), which granted them exceptional privileges, including exemption from local tithes, the right to build their chapels, and answering directly to the Pope, thereby granting them a powerful degree of independence that fueled their rapid expansion.

How the Templars secured their war steeds, and armaments was no less essential to their martial prowess. While many noble recruits inherited battle-hardened horses from family stables, the Order quickly consolidated its equine strength by establishing stud farms and supply networks within its commanderies. These warhorses cared for with the same meticulous discipline that governed all Templar operations, became symbols of freedom and relentless vigor. At the same time, master blacksmiths worked in the armories of fortified estates to forge finely wrought swords, lances, maces, and shields; each piece of weaponry was crafted with both functional brilliance and subtle religious inscriptions that testified to the divine mission behind every battle.

Beyond the glitter of arms and the gleam of polished armor, a vast logistical network underpinned the Templar enterprise. Across Europe and the Holy Land, the Order established commanderies, hospices, and fortified estates, serving as sufficient bastions that procured food, shelter, and the material means to sustain prolonged campaigns. Fertile fields yielded bountiful harvests; vineyards produced treasured wines, and well-stocked granaries guaranteed sustenance through the harshest winters and longest sieges. These centers, generously endowed through donations from noble patrons and shrewd alliances, served as administrative hubs and sanctuaries where pilgrims found respite and where the warmth of Templar hospitality alleviated the hardships of medieval life.

Integral to the fabric of the Order was the compassionate yet disciplined treatment of the people they encountered. The Templars saw themselves as protectors and benefactors, providing care, security, and justice for local communities and weary travelers. In hospices, beyond offering nourishment and shelter, they nurtured a spirit of communal solidarity. At the same time, their internal administrations, guided by a strict code of justice, ensured that Order was maintained. This balanced approach—caring for the vulnerable while meting out disciplined regulation when needed—earned the Templars a reputation as both benevolent guardians and formidable enforcers of their sacred laws.

No discussion of the early Templars would be complete without acknowledging the rigorous training and internal organization that molded them into an unparalleled fighting force. Daily life within the Order was a disciplined fusion of martial training and spiritual devotion. Knights engaged in relentless physical drills honed their horsemanship, and practiced synchronized combat tactics, all while adhering to a strict monastic schedule of prayer and reflection. Their adherence to a formalized "Rule of the Temple" further solidified the bonds of brotherhood, ensuring that every member was a warrior and a devoted servant of a higher calling. This internal cohesion, anchored in a robust hierarchical structure and guided by revered Grand Masters, was instrumental in transforming a motley band of crusaders into an elite, cohesive force whose operational efficiency would echo through the corridors of history.

Moreover, as the Order's influence expanded, so did its economic might. The Templars evolved into early pioneers of medieval banking and financial administration. They developed detailed record-keeping, secure vaults, and a vast network of wealthy benefactors, allowing them to finance their campaigns and extend their influence far beyond the battlefield. Their ability to combine spiritual fervor with a keen sense of economic strategy laid the foundation for a legacy that would continue to inspire and mystify generations.

Thus, the early Templar Order was born of a profound synthesis—a melding of divine aspiration, martial discipline, and pragmatic governance. From the hallowed grounds of Jerusalem to the sprawling commanderies across Europe, the narrative of their beginnings is one of breathtaking ambition, enduring solidarity, and ceaseless innovation. Every element—from the immaculate mantles and radiant red crosses to the meticulously managed farms, fortified refuges, well-bred steeds, masterfully forged arms, and the compassionate governance of the people—is a testament to an enduring quest for a destiny that transcends the mundane.

This rich and lavishly woven narrative beckons modern readers to delve into a saga where every battle fought, every vow sworn, and every act of mercy delivered resonates with the eternal human search for meaning—a challenge to explore a legacy where faith and power interlace in a timeless dance, echoing through the annals of history.

Chapter Two
Rise to Power & Military Influence

In the turbulent aftermath of their humble emergence, the Knights Templar transcended their origins as a modest band of crusading warriors to become an institution that reshaped the medieval world. Born in a time defined by relentless holy wars and intricate political intrigues, these devout knights harnessed divine zeal and ruthless strategic innovation. Their ascent was not accidental but the result of deliberate innovation on the battlefield, the creation of formidable fortifications, and the establishment of a vast transcontinental network that married martial might with administrative genius. The Templars' evolution from a few pious soldiers to a transnational powerhouse is etched in the echoes of clashing swords, the stone of impregnable fortresses, and the meticulous accounts of estate ledgers—a blend of spiritual fervor and pragmatic statecraft that continues to captivate history.

From the onset, the Templars distinguished themselves with unyielding piety and a mastery of military tactics that set new standards for medieval warfare. Their heavy cavalry, trained for seamless coordination, executed charges with deliberate and poetic precision—a synchronous display where every knight played an exact role in an intricate maneuver. The Order's battlefield formations were engineered to be dynamic and unyielding, enabling them to convert defensive lines into offensive thrusts with startling efficiency. However, as much as the Templars excelled in open combat, they also proved masters of siege warfare. They engineered formidable war machines—from battering rams capable of felling even the sturdiest gates to trebuchets that hurled massive stones high over the ramparts—melding functionality with symbolism, as inscriptions and heraldic emblems on each weapon linked their destructive purpose with a solemn divine mission. Crucially, their rise was significantly aided by a series of unprecedented papal bulls, notably *Omne Datum Optimum* (1139), which granted them unique independence from local bishops and the right to collect their tithes, firmly cementing their special status directly under the Pope.

A seminal moment that encapsulated this martial brilliance was the famed Battle of Montgisard in 1177. In the arid plains near Montgisard, a vastly outnumbered contingent of Crusader forces, including a distinguished contingent of Templars, faced the formidable army of Saladin. Despite numerical odds that placed the Muslim forces in the tens of thousands against a few hundred knights supported by local levies, the Templars and their allies launched a series of swift, well-coordinated cavalry charges that exploited gaps in Saladin's formation. This unexpected counterattack shattered the enemy's cohesion, forcing a hasty retreat and securing a victory that reverberated throughout Christendom. Montgisard became a legendary example of how disciplined training, innovative tactics, and sheer courage could defy overwhelming odds. This victory not only boosted morale but also cemented the reputation of the Templars as pioneers in military strategy.

While the battlefield was the stage for these displays of tactical genius, the Templars' expansion into a continental power rested on equally sophisticated recruitment and network-building methods. Though the previous chapter detailed the initial recruitment of French knights—steeped in the noble traditions of Burgundy, Normandy, and Aquitaine—this phase of expansion broadened the Order's horizons. In Italy, vibrant urban centers such as Florence, Genoa, and Rome provided steady recruits from families renowned for their advanced administrative practices and financial acumen. In regions like Lombardy and Tuscany, where centuries of banking and merchant traditions had perfected estate management and bookkeeping, Italian knights brought a refined logistical discipline that enabled the Order to manage vast estates and support protracted campaigns. In Spain, the deep martial spirit born of the Reconquista infused recruits from Castile, Aragon, and Catalonia—particularly from pivotal cities like Toledo, Valencia, and León—with an aggressive combativeness honed from centuries of intermittent war against Moorish forces. Their propensity for guerrilla tactics and rapid counterattacks injected dynamic energy into Templar warfare.

Meanwhile, in England, recruitment in counties such as Kent, Sussex, and the historic strongholds near London and Oxfordshire produced knights schooled in methodical strategy and feudal rigor. Their disciplined approach, cultivated through regional tournaments and a punctilious knightly code, provided the measured strategic oversight that balanced the reckless valor of their counterparts. This melding of French chivalric ideals with Italian financial ingenuity, Spanish martial fervor, and English strategic prudence resulted in a brotherhood capable of addressing the multifaceted demands of both combat and governance.

The Templars' shrewd political maneuvering further strengthened this diverse recruitment and established a sprawling network of commanderies and fortresses that served as military outposts and administrative centers. Strategically located along vital trade routes and near turbulent borders, these fortified centers were designed not merely as defensive positions but as nerve centers for rapid mobilization and efficient resource management. The Templars organized early systems of record-keeping and financial management within their thick stone walls, which defied numerous sieges through a combination of classical engineering and creative defensive innovations. Their commanderies functioned as hubs of power where reinvestment of revenues from agriculture, commerce, and land leases sustained the continuous construction of new fortifications and the training of fresh recruits. This extensive network also made them indispensable to the defense of the Crusader States, serving as a primary standing army and garrisoning crucial castles like Safed and Tortosa. Through early banking protocols and meticulous bookkeeping, the Order maintained an economic backbone robust enough to subsidize prolonged military campaigns and secure long-term territorial control. Moreover, their financial innovations quickly evolved beyond internal management, drawing deposits and facilitating transactions for pilgrims, nobles, and even monarchs across Europe, cementing their role as an early multinational financial institution.

The architectural legacy of the Templars stands as a testament to their integration of military and administrative vision. Their fortresses, with imposing towers, layered defenses, and labyrinthine corridors, were not only bulwarks against enemy advances but also symbols of a unified state forged on the twin pillars of faith and reason. Each fortress was an administrative stronghold in its own right, carefully positioned to oversee local estates, govern communities, and act as a secure repository for wealth and strategic assets. This network projected the Order's authority into every corner of its domain, transforming contested regions into bastions of Order and influence. While often cooperating with secular rulers and other military orders like the Hospitallers against familiar foes, the Templars' growing power also led to periods of intense rivalry and political friction within the complex dynamics of the Crusader world.

The genius of the Templar Order lay in its extraordinary ability to synthesize such seemingly disparate elements—military strategy, architectural innovation, economic acumen, and cross-cultural recruitment—into a seamless whole. Every charge executed in battle, every ingeniously constructed siege engine, every fortress that rose against the skyline, and every alliance forged with influential local powers contributed to a comprehensive system that exerted control over both the immediate battlefield and the broader realms of political and economic power. This unique commitment to spiritual purity and martial prowess distinguished them from other feudal armies, imbuing their actions with a profound moral authority and inspiring fervent support across Christendom. The celebrated victory at Montgisard resonated as a concrete example of this synthesis—a moment when well-honed tactics, the relentless spirit of a united force, and the determination to defy overwhelming odds coalesced into an achievement that transcended the limitations of its era.

Ultimately, the rise of the Templars as a military and administrative colossus was a multifaceted transformation, achieved not merely by winning battles but by establishing a legacy of innovation and governance. Their story is woven from the threads of divine inspiration and pragmatic statecraft—a tapestry that details how a visionary order integrated the strengths of different European cultures, engineered revolutionary tactics, and erected enduring fortresses that would sustain its power for centuries. As we move forward to exploring the inner workings, financial ingenuity, and spiritual rigor of the Order in subsequent chapters, the tale of the Templars in this chapter stands as a vivid reminder of how, in an age of conflict and uncertainty, visionary leadership and cross-cultural collaboration can forge an indomitable legacy that continues to inspire awe and fascination.

Chapter Three
Spires of Sacrifice: The Templars' Legacy in Jerusalem

Jerusalem was a city of eternal allure and unyielding sorrow—where stone and spirit intertwined in a tapestry woven from centuries of prayer, conquest, and loss. Its ancient walls, bathed in the golden light of dawn and shadowed by the weight of history, bore silent witness to the triumphs and heartaches of empires. When the Crusaders first breached its gates in 1099, they believed themselves anointed by destiny. Their exultant cries filled the streets, yet amid the jubilation, the stark knowledge lingered that claiming such a sacred prize would demand sacrifice beyond mortal measure.

In the uncertain years that followed, the Kingdom of Jerusalem emerged as a fragile beacon of Christian hope in a land rife with strife. Pilgrims, driven by dreams of witnessing divine relics and hallowed sites, journeyed across perilous roads where bandits and hostile armies awaited. In this turbulent milieu—a realm where faith waged war with blood—a new order was conceived. In 1119, a modest band of knights, determined to forsake wealth and nobility for a higher calling, found refuge within the venerable precincts of the Al-Aqsa Mosque. Many believed the ancient walls of that sacred place still bore the ruins of Solomon's Temple, and it was here, on the very site of the Temple Mount, that the Order established its primary headquarters, famously utilizing its subterranean vaults, known as 'Solomon's Stables,' for their horses. Thus, the Knights Templar were born. Their white mantles, emblazoned with a stark red cross, quickly symbolized unwavering resolve and grim defiance.

Each day, these unlikely warriors embodied paradox. Beneath a burning sun, they honed their skills on the dusty training fields, their clashing armor echoing the rhythm of relentless discipline. At night, in the calm, contemplative silence of stone cloisters, a young Templar might whisper quiet prayers and wonder if his sacrifice would someday be worthy of the sacred soil he defended. In such introspective moments, the roar of battle was replaced by the steady cadence of faith—a reminder that true valor was as much about inner fortitude as it was about the prowess of the sword.

As the Templars' renown spread from the courts of Europe to the scorching ramparts of Jerusalem, so too did the ambitions of their adversaries. The Muslim world, long stirred by its vigor, began to coalesce under the banner of Saladin—a commander whose blend of piety, ambition, and tactical genius inspired both reverence and dread. With Egypt and Syria unified under his rule, Saladin set his sights on reclaiming the Holy Land, determined to restore what he considered rightfully his.

The summer of 1177 witnessed destiny and chance meet on a vast, sun-soaked plain near Montgisard. Saladin, emboldened by past victories, assumed the scattered remnants of Crusader forces would crumble without resistance. However, standing against him was King Baldwin IV—frail with leprosy yet unyielding in spirit—supported by a loyal cadre of knights, including the indomitable Templars. A thunderous charge erupted under a scorching sky as sweat mingled with the clamor of clashing steel. The Templars advanced in tight formation, their lances gleaming like shafts of divine light. In that fateful moment, the battlefield transformed into an arena of miracles. Saladin's forces, caught unawares by the discipline and formidable courage of the assault, recoiled in disarray. Though fleeting in its respite, this hard-won victory rekindled hope among the Crusaders—a hope baptizing the Templars as the living heartbeat of their sacred mission. Beyond their battlefield heroics, the Templars also established a critical network of formidable castles throughout the Kingdom of Jerusalem, like Chastel Pèlerin and Safed, which served as vital defensive bulwarks, safeguarding pilgrim routes and protecting the heartland of the Crusader territories.

Nevertheless, the brilliance of Montgisard was soon eclipsed by the grim reality of war. Ten years later, in the barren plains near Hattin, summer's relentless heat drained every last reserve of strength from the Crusader host. Saladin, ever the cunning strategist, had lured them into a trap where the very air, scorching and unforgiving, served as an accomplice in their undoing. As water became a scarce luxury and armor fused to flesh under the brutal sun, the knights succumbed to despair, no matter how valiant. Amid the carnage, an anonymous Templar fought with the resolve of one who knew that each swing of his sword might be his last act of defiance. Despite their valor, the collective might of Saladin's forces proved insurmountable. In the quiet aftermath, as the dust settled and anguished cries echoed across the field, many Templars were captured and, by Saladin's merciless decree, condemned to death. The ensuing loss at Hattin would forever stain the annals of the Crusader legacy.

In the wake of Hattin, with the tides of war turning inexorably dark, Saladin turned his resolute gaze to Jerusalem. The once-proud city—now familiar with the sorrow of siege—braced for its inevitable fall. Defenders fortified every gate and rampart within its venerable walls while the anguished fragrance of burning timber intermingled with ceaseless prayers. The hammering of siege engines and the laments of a people who had once believed in resurrection filled the air until, in early October 1187, negotiations overcame the clamor of battle. That day, a quiet and sorrowful surrender was arranged. The Templars, who had valiantly participated in the city's defense, were among those compelled to lay down arms, and many were allowed to depart under Saladin's terms. Jerusalem—a city that had once resounded with the cheers of conquest—yielded to Saladin's steadfast might, marking not a triumph of victory but the deep mourning of loss.

News of Jerusalem's surrender reverberated back through Europe, stoking deep indignation and inspiring a renewed crusade. In the legendary figure of Richard the Lionheart, the Crusader spirit found a new champion. With the Templars, forever marred by the tragedies of Hattin and the fall of their cherished city, marching resolutely alongside him, the Third Crusade was set in motion. Fierce battles followed at Acre and Arsuf, where the rhythmic thunder of hooves and the brutal embrace of steel recreated the glory of earlier days for a moment. Nevertheless, even as these clashes brought temporary victories, the dream of reclaiming Jerusalem remained a ghostly yearning—a sacred promise unfulfilled. Despite the immense effort and sacrifice, the Crusaders never permanently regained Jerusalem after its fall in 1187, a lasting testament to the ultimate unfulfillment of their primary sacred mission.

As the centuries turned, the weight of history bore down on the waning Crusader states. In 1291, when Acre—the last beacon of Christian rule—finally fell, the Templars were forced to retreat from a land that had once sung with the energy of holy war. During the brutal Siege of Acre, the Templars famously mounted one of the most valiant last stands, defending their quarter with suicidal courage and suffering catastrophic losses as the city fell, marking the definitive end of their military presence in the Holy Land. They withdrew first to Cyprus and later into the intricate maze of European politics. Although their martial legacy resonated across the battlefields of the Holy Land, they gradually morphed into guardians of wealth and secret traditions, revered yet shadowed by envious hearts.

Whispers of clandestine rituals, hidden treasures, and veiled pacts soon mingled with the Templars' storied past. By 1307, King Philip IV of France, mired in debt and threatened by the considerable assets amassed by the Order, initiated a ruthless crackdown. Arrests, excruciating interrogations, and protracted trials followed until, by 1312, the once-mighty Knights Templar were formally disbanded. The crown and the church absorbed the proud bastions and secret vaults they had built, marking the formal end of an era.

However, even as official records closed on the Templars, the echo of their legacy endured in legend and lore. Tales of their exploits, whispered in hushed tones by candlelight, spoke of secret relics and mystical knowledge that had journeyed far beyond the known world—even to distant Agra.

This city, centuries later, would captivate with its splendor. Though modern historians dismiss such accounts as imaginative myths, they underscore an unyielding truth: the Templars remain an eternal enigma in the collective imagination, a testament to a time when faith and steel intertwined to shape history.

In the hushed twilight of memory, as dusk draped over the ancient stones of Jerusalem and the winds whispered lost prayers, the enduring legacy of the Knights Templar emerged as a meditation on fate, ambition, and the ephemeral nature of human endeavor. Their story is complex and interweaves luminous heroism and profound tragedy. It invites readers to reflect on how even the inexorable march of history can eclipse the noblest aspirations. In every echo of clashing swords and every shadow cast by crumbling fortifications, the spirit of the Templars endures, urging future generations to remember that even in loss, there shines a perennial light of resilience.

Chapter Four
Templar Architecture and Fortifications

Few aspects of the Knights Templar's formidable presence were as tangible and enduring as their architectural legacy. Beyond their roles as warrior-monks and financial innovators, the Templars were also master builders, erecting a vast network of strategically designed fortifications and religious structures that stretched from the sun-scorched plains of the Holy Land to the bustling cities of Europe. Their architectural endeavors were not merely acts of construction; they were precise expressions of their military might, economic prowess, and spiritual devotion, shaping the landscape of medieval Christendom in profound and lasting ways. These imposing stone testaments served as physical anchors for their vast international empire, projecting power, securing assets, and providing spiritual solace across diverse and often hostile territories, establishing benchmarks for medieval engineering and logistics.

In the crucible of the Crusader States, the Templars' architectural genius shone brightest through their formidable military fortifications. These castles were far more than simple strongholds; they were sophisticated, multi-layered defensive systems meticulously designed to withstand the relentless sieges of their formidable adversaries. Their primary function was profoundly strategic: to secure vital territories, protect pilgrimage routes traversing dangerous landscapes, serve as logistical hubs for the movement of troops, supplies, and treasure, and act as powerful, unyielding symbols of Crusader authority in a perpetually contested land. The Templars, refining lessons learned from centuries of Byzantine and Islamic military architecture, popularized and perfected the concentric castle design, a revolutionary defensive system featuring multiple rings of walls, each progressively higher than the one outside it. This ingenious tiered defense ensured that if an outer wall was breached, attackers would face another, often more formidable, barrier, all while being exposed to devastating crossfire and projectiles from the elevated inner defenses.

Key defensive features were meticulously integrated into these designs, reflecting an acute understanding of siege warfare and human psychology. Machicolations—projecting stone galleries with floor openings—allowed defenders to drop stones, boiling oil, or arrows directly onto attackers at the vulnerable base of the walls, denying them a safe approach. Steep glacis slopes, often revetted with stone, created treacherous approaches that deterred direct assaults and forced attackers into exposed positions. Barbicans and complex, winding gatehouses, sometimes featuring multiple portcullises and murder holes, were not just entrances but elaborate kill zones, funneling enemies into narrow passages where they could be assaulted from multiple angles. Towers were strategically placed at regular intervals for flanking fire, providing overlapping fields of view and defense. Hidden sally ports allowed for swift counter-attacks and raids against besieging forces, often catching them off guard. Ingenious and vast cisterns, sometimes carved deep into bedrock, ensured vital water supplies during protracted sieges, a critical factor in arid regions. Narrow arrow slits were designed to offer maximum protection to archers while providing optimal fields of fire. The Templars also demonstrated remarkable adaptation to terrain, expertly leveraging natural topography—building on precipitous hillsides, rugged coastal cliffs, or dominating strategic passes—to enhance the inherent defensive strength of their fortresses, minimizing the need for artificial defenses where nature provided them.

Among their iconic military constructions, Krak des Chevaliers stands as arguably the pinnacle of Crusader castle design, showcasing the full extent of advanced military engineering. While initially constructed by the Hospitallers, its concentric layout, formidable keeps, and intricate inner defenses were a testament to the principles shared and refined by all major military orders, rendering it virtually impregnable for nearly two centuries. Another prime example of their coastal prowess was Château Pèlerin (Atlit), a colossal fortress built directly on a rocky promontory jutting into the sea. Its strategic importance as a supply base, a naval harbor, and a staging ground for operations was matched only by its advanced defenses, which included a massive seaward wall designed to resist naval assault, deep moats carved from the rock, and a series of inner fortifications that made it nearly impossible to capture from land or sea. Other significant Templar-held strongholds like Tortosa and Safed further solidified their reputation as master military engineers, each designed to serve a critical role in the Templar defense network across the Levant. Life within these formidable structures was disciplined and arduous, a blend of military readiness and monastic routine. Knights trained constantly, their movements dictated by bells for prayer, drills, and meals, all facilitated by the castle's layout designed for efficiency and command.

While their Crusader castles were marvels of military engineering in hostile territories, the Templars' vast network of preceptories and commanderies across Europe showcased a different, yet equally impressive, facet of their architectural prowess. These complexes were the very backbone of their immense financial and administrative empire, serving as much more than just military outposts. They were intricate hubs that functioned as recruitment centers for new knights and sergeants, agricultural estates producing vital revenue, secure financial depositories for vast sums of wealth, administrative headquarters managing their extensive landholdings and vast loan networks, and spiritual retreats for their brethren.

The architectural layouts of these European commanderies often reflected a unique blend of monastic and defensive styles. While typically monastic in their internal organization—featuring essential components like chapels for worship, dormitories for communal living, refectories for meals, and chapter houses for daily governance and confession—they were almost invariably fortified. Robust perimeter walls, imposing gatehouses, and strategically placed watchtowers were common features, reflecting the need to protect their immense wealth, valuable documents, and the significant agricultural and industrial operations (mills, vineyards, workshops) that supported the order. Within these walls, one would find granaries for storing produce, extensive stables for their vast horse breeding operations, infirmaries for the sick, and sophisticated administrative buildings housing their meticulous record-keeping and financial transactions. The design facilitated a self-sufficient and secure environment, capable of operating independently and safely in a sometimes-unpredictable European landscape.

Notable examples of these European structures abound, revealing their widespread influence. The Temple Church in London, while today primarily known for its iconic circular nave and the effigies of Templar knights, was once part of a much larger, sprawling complex—the Temple Quarter—that served as the Templars' English headquarters. This site was a major banking center, a legal hub (giving rise to the modern Inns of Court), and a formidable urban fortress capable of housing a small army. Similarly, the Temple Quarter in Paris was an enormous, walled enclosure with a massive central tower (the "Tour du Temple"), functioning as the primary Templar treasury and administrative hub in France, and a symbol of their formidable power, even defying royal authority for a time. In Portugal, the Convento de Cristo in Tomar stands as a magnificent testament to their enduring legacy. Originally a Templar stronghold with its distinctive circular church, it was later inherited and vastly expanded by the Order of Christ (their successor in Portugal), showcasing a stunning evolution of architectural styles, from Romanesque to Gothic and elaborate Manueline flourishes, with its stunning Charola (round church) serving as a direct, spiritual link to the Templar past. These European structures, though often less dramatic than their Crusader castles, were equally critical to the Templars' global operations and their deep integration into the European socio-economic fabric.

The Templars' approach to religious architecture was as distinctive as their military engineering. A striking and highly symbolic feature of their sacred spaces, particularly in Europe, was the frequent construction of circular or octagonal churches and chapels. This unique architectural choice, deviating from the more common cruciform (cross-shaped) churches of the era, is widely believed to be a powerful visual and spiritual link to the Church of the Holy Sepulchre in Jerusalem, the holiest site in Christendom and the very heart of the Crusader ideal. By replicating the rotunda of the Holy Sepulchre, Templar churches aimed to bring a piece of the Holy Land's sacred power, its tangible connection to Christ's resurrection, to their European brethren and supporters, fostering a deep spiritual connection to their mission.

Beyond this direct symbolic connection, scholars and enthusiasts continue to debate whether Templar architecture contained deeper, more esoteric or mystical meanings. Discussions often revolve around the specific dimensions, numerical symbolism, and placement of architectural elements or carvings within their structures, hinting at a hidden layer of spiritual knowledge or adherence to ancient geometric principles. The Temple Church in London, for instance, with its perfectly circular nave, remains a focal point for such speculation, its design often interpreted as a microcosm of the celestial sphere, a representation of eternity, or even an echo of King Solomon's Temple. While definitive, universally accepted proof of complex esoteric codes remains elusive, the deliberate and repetitive use of these circular designs certainly set Templar chapels apart, contributing significantly to their aura of mystery and their lasting appeal to various secret societies and esoteric traditions.

The Templars' success as master builders was not solely due to innovative designs but also to their remarkable engineering prowess and logistical mastery. Constructing and maintaining such a vast and geographically dispersed network of buildings across two continents required an immense undertaking of resource management, planning, and execution unprecedented for a non-state entity in the medieval period. They efficiently managed vast quantities of raw materials, coordinating the quarrying of stone, the felling and transport of timber, and the manufacture of crucial components like iron fittings and tools. This required sophisticated supply chains stretching across land and sea.

They organized and managed large, multi-skilled workforces—comprising not only their own knight and sergeant brethren but also legions of skilled local masons, carpenters, blacksmiths, and laborers, alongside specialists brought from afar. Their expertise extended to practical construction techniques, including advanced masonry methods, the strategic use of local stone types, and sophisticated water management systems like intricate cisterns, aqueducts, and drainage systems, which were absolutely crucial for sustaining large garrisons and populations during long sieges in arid regions. The ability to source, transport, and deploy materials and labor on such a scale speaks to an exceptional level of organizational capacity and financial backing. Critically, each architectural project, from a grand Crusader castle to a remote European commandery, was not an isolated venture but an integrated part of a larger, coordinated Templar network, demonstrating their ability to plan, execute, and maintain infrastructure on an unprecedented international scale for a medieval organization. The sheer cost of these endeavors also speaks volumes about their financial might and their ability to generate and manage immense wealth.

The dramatic and brutal suppression of the Knights Templar in the early 14th century led to the rapid decline and repurposing of many of their architectural wonders. With the Order's dissolution, their immense property network across Europe and the Middle East was largely transferred to rival orders like the Knights Hospitaller by papal decree, or outright seized by monarchs. Many of their European preceptories were subsequently repurposed for new monastic orders, converted into secular residences, or simply allowed to fall into ruin, their original Templar purpose gradually fading from local memory, often leaving behind only enigmatic remnants. In the Holy Land, many of their most impressive castles eventually fell to advancing Mamluk forces, some like Atlit enduring to the very end of the Crusader presence. While formidable, even their "impregnable" designs eventually succumbed to overwhelming force, demonstrating the limits of even the most advanced medieval fortifications against determined and superior armies.

Despite this abrupt end, the Templars' lasting architectural influence is undeniable. Their innovations in concentric castle design profoundly impacted military architecture for centuries, shaping subsequent fortifications across Europe and the Levant. Their monastic designs, adapted for organizational and defensive purposes, also left a subtle imprint on later monastic and corporate structures, demonstrating how their organizational needs drove unique architectural solutions. Today, the surviving Templar buildings—whether they are the grand, weathered stones of Krak des Chevaliers in Syria, the enigmatic circular nave of the Temple Church in London, or the sprawling complex of Tomar in Portugal—stand as powerful, tangible reminders of their immense power, wealth, and sophisticated organization. These imposing stone structures continue to whisper tales of a lost order, inviting modern visitors and scholars alike to ponder the lives, beliefs, and secrets of these enigmatic warrior-monks. They are not merely historical landmarks but enduring physical links to the Templars' captivating mysteries, attracting archaeologists, historians, and seekers of hidden knowledge alike.

Chapter Five
Wealth, Banking, and Political Power

In the shadows of crusader battlefields and amid the fervor of holy wars, the Knights Templar quietly constructed an economic empire rivaling their martial exploits in both ambition and impact. While the Order had first emerged as guardians of pilgrims and warriors of the Holy Land, initially sustained by generous donations from noble patrons and widespread land grants due to their pious image and papal favor, they soon turned their attention to wealth and enterprise, establishing a sophisticated financial network that anticipated many modern banking practices. In bustling cities like Paris, London, and Barcelona, Templar preceptories were erected as fortified retreats and secure vaults and administrative centers where pilgrims, merchants, and noble patrons could deposit their valuables. By issuing letters of credit in exchange for these deposits—clever precursors to modern banking—the Templars allowed funds to be withdrawn safely at any other branch of their far-flung network, effectively reducing the risks inherent in long and treacherous journeys. Meticulous record keeping, executed on vellum using early forms of double-entry bookkeeping, ensured that every transaction was documented with precision, fueling an economy that grew to fund military campaigns, constructing impregnable fortresses, and expanding their territorial holdings.

This emerging financial prowess was intricately connected with the vast array of estate holdings spread across the European continent. The Templars secured extensive lands in regions such as Champagne, Aquitaine, the Loire Valley, and Loir-et-Cher in France. These estates, including vineyards, orchards, and manor houses, produced a steady supply of agricultural wealth and generated revenue through tolls and customs, thereby subsidizing the Order's military endeavors and infrastructure projects. On the Italian Peninsula, the Order capitalized on centuries of mercantile tradition and administrative skill refined in cities like Florence, Genoa, and Rome. Estates in Lombardy and Tuscany were managed with unparalleled efficiency, reflecting a deep-rooted understanding of property valuation and resource management that allowed profits to be reinvested in new fortifications, trade ventures, and further estate acquisitions. In the turbulent expanse of Spain, where the Reconquista had instilled a fierce martial spirit, Templar holdings in strategic locales such as Toledo, Valencia, and León offered economic rewards. They provided vital military bases to control key trade routes. In England, the Templars solidified their influence through properties in counties like Kent and Sussex and at renowned sites like the Temple Church in London, where their estates functioned as centers for spiritual devotion and economic activity, embedding the Order within the fabric of local governance and commerce.

The intricate financial administration system established by the Templars was built on a foundation of disciplined record-keeping and innovative accounting. Their painstakingly maintained and updated ledgers reveal administrative sophistication that allowed the Order to manage vast amounts of capital efficiently. This system ensured that deposits and withdrawals were securely tracked and provided a framework for internal audits and reinvestment strategies. Beyond letters of credit, the Templars commonly offered secure safekeeping for precious items, provided loans to kings and nobles against collateral, and even acted as agents for collecting taxes and papal revenues, further expanding their financial reach and influence. By channeling the wealth generated from donations, tithes, and fees for protective services, the Templars could finance everything from constructing new preceptories to providing state-of-the-art military equipment. In turn, this economic stability supported a far-reaching network that stabilized the local economy, thus stimulating agricultural development and urban growth in many regions. It also contributed to broader shifts in European financial practice, influencing the emergence of Renaissance banking and modern fiscal policy.

However, such expansive wealth and independent financial authority inevitably attracted the attention and envy of Europe's ruling elite. Monarchs and feudal lords, whose power traditionally depended on localized control of resources and revenues, began to view the Templars' burgeoning economic empire as a direct threat to their supremacy. Nowhere was this tension more acute than in France, where King Philip IV, burdened by crushing debts from his expensive wars with England and Flanders and desperate for funds, saw in the Templars an irresistible target. In a dramatic and well-orchestrated campaign in 1307, the French crown arrested Templar knights on charges ranging from heresy to financial impropriety, seizing their assets and dismantling their decentralized network to re-centralize power. Similar episodes unfolded elsewhere in Europe, where local authorities maneuvered to appropriate Templar riches, thereby igniting a prolonged conflict between the decentralized financial might of the Order and the emerging centralized state apparatus. This clash between visionary fiscal innovation and the traditional hierarchies of power marked a turning point, ultimately contributing to the Order's downfall and the absorption of their wealth into the crown's coffers.

Despite this dramatic suppression, the legacy of the Templars in wealth, banking, and political influence has endured. Their pioneering practices—secure wealth transfer through preceptories, rigorous record keeping on vellum, and the strategic reinvestment of income from expansive estates—established a blueprint that would later underpin modern financial institutions. The ripple effects of their innovations can be traced to the intricate banking systems of Renaissance Italy and the structured economic policies of early modern Europe. Furthermore, by facilitating the secure transfer of funds over vast distances, their system significantly boosted long-distance trade and pilgrimage, acting as a powerful engine for wider European economic growth. By blending sacred duty with fiscal astuteness, the Templars cultivated an aura of mystery and reverence that bolstered public trust in their financial dealings and inspired countless legends and conspiracy theories in the following centuries.

In reflecting on the Templars' financial and political achievements, one witnesses an extraordinary fusion of martial courage and administrative brilliance. Their ability to organize a transnational network of secure vaults, manage vast tracts of land, and innovate early forms of banking speaks to a visionary approach that challenged the established Order and redefined the distribution of power in medieval society. Even as their eventual suppression by monarchs highlighted the inherent tensions between decentralized private wealth and centralized state authority, the methods they pioneered resonated across time, influencing modern economic, legal, and administrative frameworks.

Ultimately, the saga of the Knights Templar in wealth, banking, and political power is a testament to the enduring potency of human ingenuity. It is a story of visionary enterprise cradled within the treacherous dynamics of medieval politics. It is a narrative where sacred duty, fiscal acumen, and an unyielding quest for influence converged to create a legacy that still inspires awe. As we turn the page to explore further facets of the Templar Order—their internal organization, spiritual ethos, and the myths that continue to swirl around their faded banners—we are reminded that their story is not only one of military conquest but also an enduring blueprint for the integration of financial prowess with statecraft. This legacy quietly reshaped the modern world.

Chapter Six
Secret Rites, Symbols & Allegations

From shadowed corridors deep within ancient abbeys to the timeworn ramparts of medieval fortresses, secret rituals once echoed through hallowed halls steeped in timeless mystery and myth. The Knights Templar—renowned for their fierce martial prowess, clandestine devotion, and profound spiritual enigma—crafted ceremonies far beyond mere formality. These rites were transformative experiences designed to shatter the boundaries of the mundane and ignite an inner flame that melded physical valor with celestial insight. Every whispered vow, every intricate gesture, and every meticulously rendered symbol beckoned the initiate to forsake the familiar and step into a realm of transcendent purpose. Envision the moment of crossing that threshold—a grand, cavernous sanctuary whose vaulted ceilings, smoothed by centuries of ritual reverence, stretch upward into dim, ancient light. The walls, rugged yet resplendent, bear carved bas-reliefs and arcane symbols recounting heroic sagas, sacred geometries, and the secret language of the divine. Here, the air is redolent with a heady blend of incense, burning myrrh, and the subtle musk of aged parchment, as if each molecule has absorbed the wisdom and sorrow of countless past ceremonies. Flickering candles cast dancing, elongated shadows across every surface, inviting the seeker to an opulently tangible and hauntingly spectral sensory banquet. Within these monumental spaces, the Templar initiations unfolded like epic dramas—initiates—often barefoot and trembling with both trepidation and fervor—traversed labyrinthine passageways by torchlight. Resonant chants and the murmuring echoes of solemn oaths punctuated each measured step on cold, polished stone. Talismans and tests were glistening ceremonial swords, iron keys, and intricately wrought crosses. In these intense moments, the old self was sacrificed upon an altar of mystery, and a reborn spirit emerged—sanctified, enlightened, and bound irrevocably to a brotherhood whose unity was forged in secret revelation.

At the core of these transformative ceremonies lay a resplendent symbolic language—a lexicon of emblems that transcended time. The Red Cross, once a stark emblem of martyrdom and crusading zeal, was reimagined by the Templars as a vibrant sign of sacrificial commitment and inner illumination. It signified the bloodshed of battle and a radiant promise of spiritual rebirth. Equally arresting was the enigmatic Beauseant banner; its striking black-and-white halves symbolized their dual nature: fair and favorable to friends but black and terrible to enemies. Its name, evoking "beautiful signs," hinted at clandestine allegiances and mystical doctrines known only to the initiated. The Templars adopted a dazzling array of cross variations to expand this visual language. The cross pattée—with its boldly outward-flaring arms—signified martial vigor and resolute spiritual fortitude; the eight-pointed Maltese cross embodied eternal rebirth and harmonious completeness; the perfectly symmetrical Greek cross spoke of balanced unity; the Jerusalem cross, with a central emblem encircled by four smaller crosses, evoked divine expansion and eternal protection; the interlaced Celtic cross recalled the endless cycle of life and death; and the austere Tau cross, with its stark "T" shape—signified that pivotal moment when destiny demanded transformation. Each emblem, borne of masterful craftsmanship and layered with sacred meaning, testified to the Templars' ability to intertwine the pragmatic with the mystical.

However, as luminous as these genuinely secret rites and symbols were, so did dark shadows of controversy and scandal embroil the order. While the Templars' internal ceremonies were likely orthodox, albeit secretive, in the medieval context, any form of secrecy could be exploited. Amid the haze of candlelight and whispered oaths, explosive allegations emerged that would tarnish the Templars' exalted image. In the tumultuous twilight preceding their dramatic downfall, influential figures—notably King Philip IV of France—orchestrated denunciation campaigns. Medieval chroniclers steeped in doctrinal rigidity leveled damning charges against the order. They claimed the Templars had perverted the sacred, that the revered red cross had been transformed into a symbol of debauchery, and that hidden ceremonies involved scandalous inversions of religious tenets. Torture-induced testimonies spoke of ritualistic subversions, acts so profane that they were said to border on the unholy, including the alleged veneration of demonic idols, an image later mythologized as "Baphomet." The term "Baphomet" itself likely originated as a medieval French mispronunciation of "Mahomet" (Muhammad), a strategic effort to link the Templars' alleged heresy to the idolatry of Islam's prophet. Such vivid, politically charged accusations—molded by coercion and rival zeal—were intended to discredit an order that combined spiritual supremacy with formidable worldly power. Accusations of denying Christ and spitting on the cross were, moreover, standard charges used against accused heretics of the era, further revealing the manipulative nature of the trials.

The scandal was compounded by lurid rumors of clandestine sexual rites that defied the moral sensibilities of the age. Sensationalized accounts alleged that, behind closed doors, initiates were forced into degrading acts that blurred, and sometimes obliterated, the boundaries between the sacred and the profane. Tales of provocative carnal excess, interspersed with sacrilegious displays against once-inviolate symbols, spread like wildfire. Although many such confessions were later retracted or dismissed as the tortured byproducts of coercion, their enduring echo continues to kindle debate, fueling conspiracy theories and artistic reinterpretations and forever imbuing the Templar mythos with a dark, enigmatic allure.

Interwoven with these grand narratives are intimate glimpses into the Templar legacy. While no extensive firsthand diaries or personal letters from Templar initiates have definitively survived history's passage, our understanding of their inner world is gleaned from interpretations of their Rule, judicial records (often extracted under duress), and later romanticized accounts. One such account recounts a young initiate, bathed in the trembling glow of candlelight, staring in awe at timeworn inscriptions that seemed to pulse with the heartbeat of ancient mysteries. At that moment—where profound fear intermingled with awe—a transformation occurred: a timid acolyte was reborn as a keeper of hidden truths. These narratives, steeped in reverence and resilience, ground the lofty symbolism in the tactile and emotional reality of "the human experience."

Moreover, the Templars' rites are part of a vast, universal continuum of secret initiatory practices. When one examines the mystery schools of ancient Egypt, the cryptic ceremonies of the Eleusinian Mysteries in Greece, or the revelatory initiations found in many Eastern traditions, a striking commonality emerges: all share the deliberate shedding of the old self, rigorous purification by elemental forces, and the rebirth into an enlightened, unified state. This unbroken chain of human aspiration affirms the timeless nature of the Templar journey and situates it within a broader, global tapestry of spiritual transcendence.

Interlinked with these transformative and mystical aspects was the raw engine of power. The Templars were equally adept as financial strategists and military tacticians as they were in esoteric pursuits. Their secret ceremonies served a dual purpose: transforming individual souls while cementing the interdependent bonds underpinning a vast network of wealth, influence, and military prowess. In an era of precarious alliances and ruthless rivalries, each ritual became a calculated assertion of discipline and unity—a demonstration intended to inspire loyalty, deter adversaries, and justify the vast accumulation of power. The fusion of sensational allegations, politically motivated confessions, and strategic economic maneuvering ultimately sealed the Templars' fate, transforming their once-revered symbols into a rich tapestry of political myth and enduring legend.

A vibrant timeline further illuminates this evolution. In the 12th and 13th centuries, the Templars laid the foundations of their mystic rites and adopted potent symbols—the red cross, the cross pattée, and more—which became the bedrock of their spiritual identity. As the 14th and 15th centuries dawned, escalating internal strife and mounting external pressures paved the way for scandalous rumors and politically charged confessions. Each accusation of heresy, each whispered tale of debauchery, served as an instrument of subversion and a catalyst for transformation. The Renaissance sparked a revival of esoteric inquiry, as occult scholars and emerging secret societies eagerly appropriated Templar imagery, reassembling these ancient icons for new, philosophical and mystical explorations. In the modern era, the Templar legacy thrives across diverse media—from digital art and blockbuster narratives to the clandestine rituals of contemporary secret orders—all of which draw on an ever-relevant mythos of resistance, renewal, and the unyielding quest for transcendence.

Finally, the sensory richness of the Templar world deserves special emphasis. Envision a majestic chamber where every architectural detail—from intricately carved vaults to the interplay of flickering candlelight with ancient stone—has been meticulously orchestrated to evoke awe and introspection. Sumptuous altars draped in deep crimson velvet, accented with shimmering golden filigree, stand alongside relics that exude historical gravitas and profound spiritual allure. The ambient sounds, whether the gentle rustle of ancient tapestries or the soft murmur of whispered prayers, combine with rich aromas of incense, myrrh, and musty parchment to create an immersive environment that engages every sense. In such an entrancing setting, the environment becomes an active participant in the ritual, a living canvas that imbues the entire ceremony with an ineffable, transcendental energy.

In the echoing silence of these hallowed spaces—where every whispered legend and every controversial allegation coalesces—the Templars invite the modern seeker to embark upon an eternal journey of discovery. Their secret rites, steeped in luminous symbolism and dark controversy, challenge us to peer beyond surface narratives and embrace the transformative power of hidden truths. The storied legacy of the Templars, woven from threads of divine light and shadow, remains an enduring testament to humanity's ceaseless pursuit of meaning, belonging, and the sublime mystery of existence.

Thus unfolds the sprawling epic of the Templar rites—a multilayered tapestry where every secret symbol and every incendiary controversy converge to form a timeless invitation to renewal and introspection. This comprehensive chronicle stands as an all-encompassing call to those willing to traverse history's hidden corridors and discover, amid the interplay of ritual and legend, the eternal promise of self-revelation and the transformative power of the clandestine.

Chapter Seven
King Philip IV & the Fall of the Templars

Under the dancing glow of flickering candlelight in an ancient Parisian monastery, the stone arches seemed to whisper secrets of a bygone era. Here, amid centuries-old carvings and muted echoes, the Templar Order had once stood as a pillar of chivalry, a coalition of fierce warriors and pioneering bankers. Their emblem—a radiant cross intertwined with cryptic symbols and arcane inscriptions—was far more than a mark of belonging; it was a key to hidden mysteries and long-forgotten lore. In what might have been the thoughts of a young Templar scribe, captured in later interpretations, "Our symbol is the spark of forbidden knowledge, a light to decipher truths concealed by time and treachery." Such words set the tone for a saga where steadfast honor would soon be assailed by ambition and betrayal.

At the height of their glory, the Templars transformed from mere crusaders into Masters of Finance and guardians of sacred relics. Their network stretched across Europe—linking distant monasteries, fortified castles, and secret chambers where obscure rituals were held under the silver glow of a full moon. In hushed councils, knights debated the meaning behind esoteric codes and hidden manuscripts, believed by some to contain the keys to immortality and divine favor. This intricate web of power and mystery made the Templars indispensable to monarchs and clerics. Nevertheless, it also sowed the seeds of envy and suspicion among those who yearned to control such extraordinary influence.

Amid this glimmer of ancient power, King Philip IV of France emerged—a monarch tormented by fiscal woes and an insatiable thirst for absolute rule. In the private solitude of his candlelit study, Philip's eyes would wander over detailed maps and secret notes scrawled in the margins. His journals, which modern historians interpret from his actions and pronouncements, reveal a man ensnared by desperation and ruthlessness that knew no bounds, filled with turbulent declarations like, "The Templars, with their unfettered wealth and clandestine rites, are both a boon and a threat—destined to be swept away for the sake of enduring royal majesty." Plagued by insoluble debt from his costly wars with England and Flanders and a fractured nobility, he resolved to seize the treasures of the Order and, in so doing, consolidate his power over both secular and spiritual realms.

Not far from the corridors of royal ambition, Pope Clement V wrestled with his crucible of conscience. Behind the heavy oak doors of the papal palace, the pontiff poured over lengthy missives that spoke of divine duty and the crushing weight of earthly authority. Torn between upholding the sanctity of the Church and bowing to the relentless pressure of King Philip, his private thoughts, often inferred from his correspondence, bemoaned, "How can faith remain unblemished when ambition forces even the holiest of us to commit unholy acts?" This inner turmoil would soon find expression in decisions that forever altered the fate of the Templars.

Then came that fateful day, Friday, October 13, 1307—when destiny turned its dark eye upon the Order. As a pall of dread descended, royal soldiers moved like shadows, breaching the hallowed sanctuaries of the Templars with brutal precision. In an account that would echo from the period, perhaps a recollection filtered through the harsh judicial process, "The furious knock at our door signaled more than the arrival of men—it heralded the end of an era. In one heartbeat, trust was shattered, and the light of our Order flickered on the verge of extinguishment." Under the weight of betrayal, loyal knights were dragged from their sanctuaries as whispers of heresy and blasphemy—fabricated to suit royal whims—filled the corridors of power. These fabricated charges included the denial of Christ and spitting on the cross during initiation, the alleged worship of an idol called Baphomet, accusations of sodomy, and the usurpation of priestly powers through mutual absolution—all standard tools in the medieval inquisitorial arsenal designed to ensure conviction.

Within the damp, oppressive dungeons, the true horror of the purge unfolded. The instruments of torment—iron racks, merciless scourges, and the searing bite of torchlight—became the grim arbiters of justice. Many Templars, renowned for their stoic valor, endured unspeakable pain rather than reveal the secrets of their Order. However, the suffering forced false confessions from others, and their voices were reduced to desperate whispers that stained their honor. Even as their anguished cries echoed off cold, stone walls, the cryptic motifs etched into these walls hinted at a resilience of spirit—a subtle defiance encoded in every chiseled rune.

In a final act that seared the world's memory, the Order's fate was sealed in a somber clearing under a bruised sky. Jacques de Molay, the last Grand

Master, stood dignified before a towering pyre, his eyes reflecting sorrow and unyielding defiance. As the flames lapped at his form, he lifted his voice to pronounce a curse—a vow against his betrayers: "May the fires you kindle consume you, and may your souls be ensnared by the sins you have wrought!" Witnesses later murmured of mysterious retribution, for soon after his execution, both King Philip IV and Pope Clement V met untimely demises as though the curse itself had returned upon those who had forsaken justice and honor. This dramatic execution was followed in 1312 by Pope Clement V's formal dissolution of the Order through the papal bull *Vox in excelso* at the Council of Vienne, a decision made under immense royal pressure.

With de Molay's passing, the majestic tapestry of the Templar Order unraveled completely. The once-powerful knights were dispersed into the obscurity of exile. While Pope Clement V officially transferred the Templars' vast wealth and extensive lands to the Knights Hospitaller, King Philip IV still managed to appropriate significant assets through various means, including demanding hefty "recovery costs" and forcing the Hospitallers to purchase the confiscated Templar properties at inflated prices, thereby filling the coffers of his crown. This vast confiscation rippled throughout medieval Europe, upending established financial practices and catalyzing the emergence of new secretive societies. Among these were the proto-Freemasons—groups that, according to later traditions and much historical debate, covertly adopted Templar symbols and ritualistic traditions, ensuring that the Order's legacy would persist in whispered legends and clandestine meetings for centuries.

The mystery surrounding the Templars grew more enchanting after their dissolution. Modern scholars, conspiracy theorists, and storytellers have pored over every surviving inscription, debated every hidden passage in ancient fortresses, and speculated on the possibility of undiscovered treasure troves. Archaeological discoveries in forgotten monasteries and recitations of coded prayers in long-forgotten manuscripts continue to stoke the flames of fascination, blurring the lines between verifiable history and the perennial allure of myth.

Thus, King Philip IV's narrative and the Templars' fall are a timeless testament to the interplay of power and fate. It invites us to ponder the recurring themes of ambition, betrayal, and the indomitable pull of secret knowledge. Whether it is the personal anguish captured in what might have been a young squire's diary, the calculated ambition seared into the private confessions of a desperate king, or the enduring curse pronounced by a dying Grand Master, this saga compels us to question: How do symbols of power persist long after empires crumble? Moreover, in the delicate interplay of fact and folklore, might we yet discover a piece of our hidden truths?

Chapter Eight
The Escape & Survival Theories

In the aftermath of the Templars' cataclysmic downfall—when the anguished echoes of burning crosses and shattered oaths threatened to fade into oblivion—a chorus of whispered legends arose along the fringes of history. While historical consensus marks the Order's official disbandment, persistent theories and romanticized narratives suggest that some Templars, defying the royal purge, managed to escape, carrying clandestine gatherings and secret legacies. At the center of these stories often rises Gérard de Villiers—not a solitary refugee but a name that became a symbol, in legend, of defiance and continuity. According to these accounts, the de Villiers family, long entwined with the fate of the Templars, emerged as the silent custodians of arcane manuscripts, cryptic genealogies, and oral traditions that chronicled every twist of a desperate exodus.

Even as the iron fist of persecution swept across Europe, disparate sanctuaries are said to have begun to flourish. In remote enclaves of France—among isolated abbeys and weathered hamlets encircled by ancient forests—fleeing Templars and de Villiers descendants allegedly found a haven. Here, secret congregations reportedly formed under the cloak of night. Family crests bearing subtle Templar insignias were exchanged in hushed, whispered meetings, and cryptic maps were drawn to lead them to further refuges. These sanctuaries provided protection and allowed once-forgotten genealogies to intertwine with local nobility; the blood of ancient knights was echoed in the heraldry of families that still bore secret symbols of honor.

In the wild, windswept highlands of Scotland—where mists weave through rugged glens and ancient stone circles stand as silent sentinels—the legacy of the Templars was reimagined in popular lore. There, the de Villiers lineage, fused with the Celtic spirit, acquired a new mystique. Secret societies are said to have flourished in secluded glens, where indigenous clans welcomed these noble exiles. In these hidden gatherings beneath brooding skies, medieval codes of chivalry merged with ancient Celtic mysticism, ensuring that the honor of the knights would transform rather than vanish, reborn in a fusion of ritual and remembrance.

Nevertheless, the story of survival was far from confined to Europe. Driven by the promise of new beginnings and the instinct to evade tyranny, audacious Templars and their loyal families are theorized to have embarked on perilous sea voyages toward uncharted horizons—the New World. Legends speak of ghost ships emerging from the fog: spectral vessels laden with encrypted manuscripts, lost relics, and secret maps directing destinies toward hidden treasure caches. These mysterious ships, battered by unyielding gales, are said to have buoyed the legacy of the Templars by delivering both their tangible wealth and intangible wisdom to nascent colonial settlements. In these fledgling communities, colonial architecture itself became an archive. Beneath the arches of early American churches and within the intricate planning of New World towns, faint yet unmistakable Templar motifs are etched into stone—a silent testimony that even in exile, the spirit of the Order perseveres.

In these far-flung lands, the survival of the Templar tradition was not merely about evasion but transformation. The de Villiers family is often depicted as having intermarried with local elites, weaving their storied bloodline into the fabric of new aristocracies. Mysterious crests emblazoned with Templar emblems and carefully hidden heirlooms passed from generation to generation, silently testifying to the legacy of honor. Lost artifacts—encrypted relics inscribed with ancient prophecies and secret instructions—became coveted treasures, and their mystery deepened with every passing age.

Modern investigative techniques have added yet another dynamic layer to this timeless tale. Digital archaeologists now employ drone-based LiDAR surveys, high-resolution scanning, and computer-assisted cryptographic analysis to unveil previously inaccessible sites. While mainstream archaeology has not confirmed a widespread, organized Templar survival, recent discoveries of shadowed tunnels, forgotten crypts in European abbeys, and colonial structures adorned with cryptic inscriptions serve as tangible clues that fuel theories that the survival routes of the Templars and the covert handoffs by the de Villiers were not merely legend. Instead, these modern tools have unearthed inscriptions that proponents claim hint at encrypted prophecies and hidden alliances that continue to challenge our understanding of medieval allegiances and occult practices.

Moreover, the occult underpinnings of this saga have never waned. In dim, secluded lodges and ancient monasteries, secret meetings have taken place for centuries—gatherings where ritualistic practices, occult symbols, and esoteric rites echo the long-lost language of the Templars. These ceremonies, long preserved and adapted by revived Masonic lodges and other clandestine orders, often claim a connection with a heritage that honors both faith and forbidden knowledge. It is important to note that the historical evidence for a direct, unbroken organizational lineage between the medieval Templars and modern secret societies is widely debated and generally considered symbolic rather than factual by historians. In these settings, ritual becomes a living language bridge linking the mystical traditions of medieval knightly orders with modern pursuits of spiritual truth.

The enduring legend of escape and survival has transformed art, literature, and culture. From Renaissance masterpieces to blockbuster films and bestselling novels, the imagery of ghost ships, clandestine meetings, and hidden crypts has continued to inspire countless creative works. This vibrant cultural renaissance has not only reimagined the noble past but has also infused contemporary discourse with a renewed fascination for mystery and conspiracies—a subtle reminder that the legacy of the Templars, though shrouded in myth, retains a persistent hold on global imagination.

Political intrigue and economic conspiracies further add to this tapestry. Some accounts suggest that surviving Templars, armed with secret financial instruments and guarded treasures, became foundational figures in emerging transatlantic trade networks. Their discreet funding and subtle influence on colonial politics hint at a clandestine financial power that, even in exile, meddled with the formation of new economic elites. Archived correspondence and cryptic political pamphlets from early colonial administrations occasionally reveal references to clandestine allegiances that suggest the untouched legacy of the Templars continued to shape the economic engines of a New World.

Thus, whether sheltered in the secluded mountain refuges of France, hidden beneath the mossy glens of Scotland, or secretly embedded within the architectural marvels of colonial America, the saga of the Knights Templar's escape endures as a multifaceted epic. The de Villiers family emerges as both guardian and innovator—a lineage that not only orchestrated daring escapes but also ensured that Templar wisdom, encrypted in relics and transmitted through secret rituals, would persist across the centuries. Their legacy is etched in bloodlines, inscribed in stone, and continually rediscovered by modern eyes probing digital archives and ancient ruins.

Every element—from ghost ships haunting stormy seas to digital scans uncovering hidden passages, from the mystical symbols in colonial crypts to the modern discussions in secret societies—coalesces into one eternal affirmation: the spirit of the Knights Templar endures. It lives on in secret genealogies, whispered traditions, and the relentless modern pursuit of hidden truth—a legacy unbound by time, forever bridging the chivalric past and the digital present.

Chapter Nine
The Hidden Templar Treasure

When the mighty reign of the Knights Templar crumbled beneath the crushing weight of royal persecution, a legend was set in motion—destined to captivate treasure hunters, scholars, and dreamers for centuries. As the royal edicts fell like a guillotine on the Templars and the blazing crosses of punishment faded into the cold night, whispers circulated about a vast and secret hoard. The existence of such a grand, hidden Templar treasure, beyond what was confiscated by authorities or transferred to other orders, remains mainly in the realm of folklore and fervent speculation, yet its enduring allure is undeniable. According to these murmurs, gold coins brought from the Holy Land, sacred relics believed to have graced venerable altars, intricately crafted chalices, and other priceless artifacts were not simply seized by the crown. Instead, the legend claims, the Order, prescient in its final moments, deliberately dispersed its considerable fortune. This calculated concealment ensured that even in their ruin, the legacy of the Knights would survive—and with it, the promise of an eternal treasure, hidden away in vaults and encoded in secret texts, its very elusiveness cementing its mythical power.

In the quiet corridors of medieval monasteries and among the crumbling stone walls of ancient abbeys in rural France, a fictional account often cited in popular lore describes a faded diary was discovered. Its yellowed pages, written in a trembling hand, supposedly recorded a vivid account of that fateful evening. One entry allegedly recalls the somber departure of carts laden with glittering treasure, rolling out under the subdued glow of moonlight as if hurriedly dispatched to preserve every ounce of their wealth. The knight who penned these lines is said to have been entrusted with an encrypted map—its cipher promising to reveal the locations of hidden vaults carved deep beneath the fortress's foundations. Whether these scenes of treasure-laden carts in Paris were entirely factual or later romanticizations born from the fevered need to believe in hope, they remain one of the most evocative images associated with the Templar flight, a cornerstone of the treasure hunter's dream.

Yet the mystery extends beyond physical concealment. The Templars, renowned not only as fierce warriors and pioneering financiers but also as ingenious cryptographers, are theorized by some to have ensconced their riches in more than mere stone and mortar. In manuscripts seemingly dedicated to devotional writings, intricate codes, and encrypted instructions were interwoven with ornamental calligraphy. Modern scholars now approach these documents with computer-assisted decryption techniques, seeking the hidden coordinates and keys that might unlock concealed chambers. No widely accepted decryption of Templar texts has led to the discovery of a vast, hidden treasure. However, the idea that the actual location of this treasure is encoded within layers of symbolism, the product of a strategic mind as advanced as it was devout—adds an intellectual challenge to the hunt, one bridging the gap between myth and method, inviting endless intellectual pursuit.

Deep within the folds of Templar history, heraldic mysteries also emerge. Many noble families across Europe and even in the New World have borne crests and emblems that, upon closer inspection, proponents suggest, reveal unexpected Templar motifs. These symbols, discreetly passed down through generations, are sometimes interpreted as hinting at secret genealogies and indicate that, far from vanishing, the bloodline and the secret custodianship of Templar lore persist in quiet corners of aristocratic society. However, it is crucial to note that these genealogical links and claims of direct bloodline custodianship are primarily theoretical and not substantiated by conclusive historical evidence. Ancient family records and unexplained carvings on historic buildings contribute to an enduring debate: Are these heraldic echoes a deliberate signature left by the Knights, intended to signal their legacy to the chosen few, or merely coincidental resemblances and later fabrications?

Equally fascinating are the threads of prophecy woven through this saga. Medieval accounts speak of prophetic dreams and divine omens—a dying knight's cryptic last words foretelling the eventual rediscovery of their treasure or visions recorded in secret journals that hinted at a future day when the lost hoard would restore the Order's faded glory. These prophetic elements exist primarily within medieval folklore and later romanticized histories, elevating the treasure from simple material wealth to a symbol of hope and destiny, suggesting that even heaven had a role in conspiring to preserve the secret for the rightful heirs. Such narratives reinforce the mystical dimension of the Templar legacy, implying a divine hand in its preservation and ultimate revelation.

The modern era, too, has embraced and revived this ancient mystery. Across the globe, various secret societies and revivalist groups today claim to be inheritors of the Templar legacy. In clandestine gatherings held in underground lodges and time-worn castles, enthusiasts conduct rituals and display relics purportedly connected to the Order. Interviews with self-styled "custodians" of Templar lore and reports of secret ceremonies have brought this historical enigma into the present, fusing a spirit of revival with national cultural pride and a mystical touch. Such modern revival movements serve as tribute and proof that the passion for Templar secrets endures, fueling contemporary quests for tangible evidence of a storied past.

The cultural impact of the Templar treasure is further underscored by lost correspondence and personal letters. While authenticated Templar correspondence specifically detailing treasure concealment remains elusive, fragments of secret letters, sometimes cited in popular accounts and speculative histories, detail the inner turmoil, the fierce loyalty, and the solemn vows made by Templar knights as they prepared to safeguard their legacy at any cost. These poignant documents offer a human dimension to the legend, transforming the treasure into not merely a stockpile of wealth but an emblem of sacrifice and an enduring promise to a higher calling.

Adding to the mystique, reports of paranormal phenomena have been recorded at several purported Templar sites. Locals in modest Normandy villages, for instance, speak of a "golden light" that appears in remote vaults on moonlit nights, while elders in Scotland recount eerie apparitions near ancient stone circles. Such spectral encounters, often documented by casual observers and dedicated paranormal investigators, suggest that a ghostly custodianship still watches over these secret hoards, a metaphysical extension of the Knights' enduring vigilance. These tales, firmly rooted in local folklore, contribute significantly to the treasure's ethereal and captivating aura.

Beyond traditional narrative, modern digital innovation has transformed the search into an interactive journey. Imagine an augmented reality experience that allows readers to virtually traverse the hidden vaults and labyrinthine corridors once guarded by the Templars. Interactive maps with digitally reconstructed images of secret passageways and cryptic inscriptions offer the chance to decode clues in real-time. This fusion of technological advancement and historical mystery enthralls today's audience, turning the hunt into an immersive, participatory adventure.

Finally, recent scientific breakthroughs have breathed fresh life into this enduring quest. Drone-mounted LiDAR surveys and ground-penetrating radar have uncovered anomalies beneath familiar medieval sites, and meticulous forensic analysis of aged manuscripts has revealed patterns that may serve as clues to long-hidden vaults. While no definitive, large-scale Templar treasure hoard has been unearthed or confirmed by these methods, these scientific endeavors, detailed in recent archaeological reports, reinforce the argument that the Templars' methods for hiding their wealth were as innovative as they were deliberate. What was once considered solely the realm of legend now increasingly appears as an archaeological mystery waiting to be unraveled by the combined efforts of historians, scientists, and intrepid explorers, leaving open the tantalizing possibility that scientific rigor might one day bridge the gap between myth and verifiable truth.

In synthesis, the legend of the hidden Templar treasure is a breathtaking tapestry woven from threads of myth and material wealth, intellectual ingenuity, and spectral guardianship. It encapsulates the Templars' calculated dispersal of riches, the enigmatic reports of treasure-laden carts departing Paris, the sophisticated encryption of secret texts, and the eternal hope symbolized by noble emblems and prophecies. It is a mystery that bridges medieval valor and modern technological inquiry. This legacy continues to challenge and inspire those with the courage to search for truth in the hidden recesses of history

Chapter Ten
The Fleet of La Rochelle & Possible Exile

When France's iron grip descended upon the Knights Templar, centuries of chivalric power were suddenly under siege. Nevertheless, amid royal decrees and the thunder of impending doom, the port of La Rochelle emerged as an unlikely stage for an extraordinary maritime escape. In those fraught final moments, as orders were issued with ruthless finality and the Templar order was marked for extinction, secret whispers told of ships laden with treasure and hope slipping away under the cloak of twilight. It was said that, instead of allowing the crown to plunder their riches, the Templars risked everything to disperse their wealth and the essence of their legacy upon the mercurial seas.

Beneath a sky heavy with storm clouds, the atmosphere on the docks of La Rochelle was one of frantic urgency and sorrow tempered by resolve. Eyewitness testimonies passed down through generations of local mariners depict a scene where dozens of vessels—ranging from swift galleys to stalwart coastal ships—were hurriedly prepared and set afloat amid the swirling fog. A vivid account, often cited in popular lore as a journal entry by a Templar captain captures that night in poignant detail. However, its direct historical provenance remains unconfirmed: "The chill of the sea and the taste of salt carried our final prayers. Every man felt the weight of destiny as we loaded relics, secret scrolls, and the last of our gold. With the royal fleet nipping at our heels, our only chance was silence and swift departure. The darkness became our ally, and we vanished into the unknown, trusting the sea to hide our hopes and history." Such personal accounts bring texture and heart to the dramatic exodus, inviting readers to experience the strategic genius of the escape and the raw emotion of men facing the end of an era.

Scholars have long debated the fleet's intended destination. One theory holds that these treasure-laden vessels charted a desperate course north toward Scotland's rugged, mist-shrouded coasts. There, amid the timeless beauty of the Highlands, with its secret coves and ancient stones bearing inscriptions of lost lore, local traditions whisper of hidden sanctuaries where exiled Templars might have found refuge among the Celtic clans. In remote Scottish villages, elders recall eerie nights—with ghostly illuminations dancing on the water's edge and mysterious symbols carved unexpectedly into cliff faces—that seem to echo the elusive presence of a maritime exodus.

Other voices point to Portugal as the likely refuge. With its legendary status as a seafaring nation, fortified harbors, and narrow, winding streets filled with history, Portugal provided an opportune environment where the exiled Templars could rebuild their scattered community. In secret gatherings recorded in coded letters and private documents of noble families, mysterious meetings were held in dimly lit taverns along Lisbon's back alleys. These accounts hint at covert alliances between local patrons and

Templar refugees, suggesting that fragments of the order's treasure—and its encoded wisdom—were absorbed into a new land's cultural and economic fabric.

Adding to the intrigue, maritime legends lend the story to a spectral dimension. Throughout the Atlantic coasts, fishermen and local lore keep alive reports of ghost ships—a spectral fleet that sometimes appears amid thick fog, their outlines wreathed in an uncanny luminescence. Whether interpreted as omens or the restless spirits of Templar knights eternally guarding their reclaimed treasure, these supernatural sightings blur the line between legend and history. They serve as a timeless reminder that while the fleet may have vanished from mortal eyes, its legacy endures in the whispered accounts of the living.

Modern technology, too, has begun shedding light on this age-old mystery. Advances in maritime archaeology have introduced drone-mounted LiDAR imaging, ground-penetrating radar, and deep-water sonar mapping to the search. While no definitive Templar fleet wrecks or treasure ships have been unequivocally identified to date, recent underwater surveys conducted near La Rochelle and along the coasts of Scotland and Portugal have revealed anomalies—submerged structures, remnants of medieval ship timbers, and unusual geometric formations that defy natural processes. These scientific breakthroughs, detailed in contemporary articles published in the Journal of Archaeological Science and related fields, offer tantalizing clues that reaffirm the possibility of a grand maritime escape.

Beyond these technological and speculative threads lies a story of cultural survival. Even as the Templars slipped away by sea, scattered communities of survivors began to form in secret. Hidden cells in remote Scottish hamlets and Portuguese coastal towns continued to exchange encrypted correspondences, safeguard relics, and pass on the sacred codes and traditions of a once-majestic order. In these shadowed enclaves, the exiled Templars thus preserved their material wealth and the indomitable spirit that had defined their identity.

The saga of the Fleet of La Rochelle and the subsequent exile of the Templars is a multifaceted epic—an interweaving of history and myth, material triumph, and spiritual endurance. It is a narrative that captures the tangible evidence of escape—a fleet disappearing into the Atlantic—and the intangible legacy of an order that dared defy its fate. Whether the exiles found sanctuary among the stormy coasts of Scotland, within the storied harbors of Portugal, or even vanished into the expanse of the Atlantic, their enduring legacy challenges us to explore the mysteries of faith, resilience, and survival.

In the interplay of these threads—of whispered diary entries and cutting-edge digital reconstructions, of ghostly apparitions and secretive, clandestine networks—the legacy of the Templars emerges as something greater than mere treasure. It is a bridge between a chivalric past and a modern age of discovery that even when an empire falls, its spirit can endure hidden shores, in cryptic texts, and in the hearts of those bold enough to seek the truth.

Chapter Eleven
The Oak Island Mystery

In the early decades of the 12th century, a small band of nine knights set forth on what would become a sacred mission echoing down the corridors of history. In 1119 AD, driven by a fervent desire to protect pilgrims traversing the perilous routes of the Holy Land, these warriors founded the Order later known as the Knights Templar. Initially tasked with defending the weak against marauding bands and hostile armies, the Templars quickly became an elite fighting force. Equally remarkable was their groundbreaking role in early international finance. By establishing a system of secure, interlocking networks that allowed monarchs to deposit vast treasures in centers such as Paris and later safely withdraw funds in distant Jerusalem, the Templars amassed wealth and influence that would eventually become woven into legend. Their meticulously crafted seals and strictly enforced codes of conduct laid the groundwork for a public image of a formidable Order. Nevertheless, the later, often romanticized, interpretations of their symbols would inspire treasure hunters and scholars for centuries to come.

Even in those early days, the reach of the Templars extended far beyond the immediate realms of warfare and commerce. Rumors and later speculative theories abounded that these knights covertly undertook excavations beneath crumbling stables and abandoned corridors while stationed near ancient ruins allegedly related to the lost Temple of Solomon. Proponents of these theories suggest they were not merely seeking shelter but were driven by a deeper quest for sacred artifacts or hidden knowledge, perhaps even unearthing long-lost texts or the legendary Ark of the Covenant. Chroniclers from later eras and modern proponents of the Grail legend claim that during these secret forays, they discovered cryptic clues—arcane markers purportedly hinting at the location of an artifact that exhibited immeasurable power: the Holy Grail. Such early experiments in esoteric archaeology and the art of coded symbolism would ultimately infuse the Order's legacy with an element of mystery that transcended mortal affairs. In later centuries, this very tradition of guarded secrets would ignite the imaginations of adventurers who believed that similar methodical techniques might have been employed to design a treasure map allegedly etched into the natural landscape of a remote island that now goes by the name Oak Island.

As the Templars' renown spread throughout medieval Christendom, the Order came to be seen not only as formidable warriors and ingenious bankers but also as custodians of secret knowledge. Their financial innovations and cryptic emblems were far from arbitrary; every carefully wrought cross and every scripted seal formed an integral piece of coded language meant to secure a legacy for those deemed worthy. This guarded tradition of clandestine communication would ultimately prompt modern treasure hunters and researchers to wonder whether those very methods might have been employed to construct a hidden map, one that has, over time, been lost in the mists—only to be rediscovered on Oak Island.

Nevertheless, as is often the case in human affairs, fate proved merciless. In 1307, the wealth and influence that had elevated the Templars to legendary status became tools of their undoing. King Philip IV of France—burdened by astronomical debts to the Order and stirred by ruthless ambition—set in motion a calculated purge. On Friday, October 13, 1307, a date that would forever be associated with ill omens, Philip ordered the mass arrest of Templar leaders. Charges of heresy, corruption, and moral degeneration were leveled against these once-revered protectors in a public spectacle of cruelty and betrayal. In the ensuing weeks, under the crushing weight of royal authority, Pope Clement V issued a decree on November 22, 1307, which mandated the suppression of the entire Order. Amid the torments and public executions that followed, clandestine whispers emerged: some Templars, it was said, had managed a desperate escape—and in doing so, they safeguarded sacred relics (perhaps even the fabled Holy Grail) from falling into the hands of their oppressors. Many of these survivors would eventually find refuge in far-flung regions, with Scotland emerging as one storied sanctuary for those in exile.

In Scotland's rugged, mist-veiled landscapes, exiled Templars intermingled with the local aristocracy, sowing the seeds of a legacy that continues to resonate in the centuries since. In the wild expanses of the Scottish Highlands, the influence of the medieval Order became interwoven with indigenous Celtic mythology. Legendary families such as the Sinclairs—whose ancient bloodlines are steeped in whispered Templar connections—emerged as custodians of secret traditions. Over time, these families transformed mystical lore into tangible symbols, manifesting in monuments like Rosslyn Chapel. Nestled amid ancient woodlands and perpetually shrouded in myth, Rosslyn Chapel remains celebrated for its labyrinthine stone carvings and enigmatic, embedded symbols. Modern researchers who have scrutinized the chapel's intricate details note that its hidden recesses and carefully proportioned motifs appear designed to encode secret messages—visual echoes of the Templar's guarded language. In every chiseled figure, there is the enduring testimony of an age-old language of symbols, challenging the modern observer to decode the long-lost legacy of those exiled knights.

As centuries passed to the dawn of the Renaissance, the intellectual climate of Europe underwent a dramatic transformation. In the wake of medieval darkness surfaced a surging wave of reform, inquiry, and rediscovery that would irrevocably alter the world. During this transformative period, Sir Francis Bacon emerged as one of the era's most influential luminaries—a thinker whose groundbreaking contributions to philosophy and the scientific method would soon be joined by his persistent fascination with ciphers and secret codes. Born in the mid-1500s, Bacon was a man of immense curiosity. His treatises overflowed with allusions to hidden orders and clandestine messages, suggesting that layers of meaning are waiting to be uncovered beneath the veneer of everyday existence. Bacon's passionate followers—the Baconians—advanced such radical theories that, in their view, Bacon's true genius was demonstrated through his ability to encode profound truths for posterity. Some even contended that he was the elusive force behind works traditionally attributed to Shakespeare. The noticeable absence of original manuscripts only deepened speculation that these precious documents had been deliberately concealed in secret vaults, perhaps even beneath the enigmatic soil of Oak Island, as some ardent Baconians propose. Bacon's memorable assertion that he would one day be "known for who he is long after his death" acquired an almost prophetic quality, fueling the idea that his hidden legacy was meant to endure in a series of coded prophecies that only the most persistent would eventually unravel.

Bacon's explorations were not confined solely to theoretical realms; he delved into practical solutions for preserving knowledge. In his experiments, Bacon used mercury to protect delicate texts from the corrosive passage of time. Later, during early excavations on Oak Island, treasure hunters uncovered empty flasks that bore traces of mercury—a detail which, within the context of these theories, many today interpret as a deliberate echo of Bacon's preservation techniques. For those who see a continuity between the secretive practices of the Templars and the intellectual pursuits of the Renaissance, this discovery is anything but coincidental; it is viewed as a vital link in a chain of encrypted wisdom that stretches from medieval battlefields to modern treasure hunts.

Not long after Bacon's influential years, another remarkable character emerged: Thomas Bushell. Born around 1593, Bushell began his life in humble service to Bacon but quickly distinguished himself as a master miner and engineer. His pioneering work in salvaging treasures from flooded mines—conducted with the assistance of experienced Cornish miners—revealed innovative methods for water management and subterranean design. These techniques, proponents argue, share an uncanny resemblance to the specialized engineering of Oak Island's fabled Money Pit. Bushell's ingenuity eventually earned him significant responsibilities, including administration of a Royal Mint in Wales—a role that demanded exceptional accuracy in handling the kingdom's wealth.

Furthermore, during the turbulent period of the English Civil War, Bushell became renowned for his valiant defense of Lundy Island in the storm-beaten channels of the Bristol Channel. Although his prominence waned briefly after Lundy's surrender in 1647, his swift historical reemergence deepened the enigma surrounding his technical prowess. Modern engineers and historians who subscribe to these theories often compare Bushell's inventive methodologies and the sophisticated designs unearthed on Oak Island, suggesting that core principles in flood control and decoy construction were refined and transmitted across generations.

When the 18th century emerged, the narrative of hidden treasure took a decidedly maritime turn. European powers, driven by imperial ambition and the lure of uncharted riches, embarked on bold voyages to reshape global history. In its relentless quest to secure dominion over the New World, France undertook an ambitious project: constructing the formidable Fortress of Louisbourg on Cape Breton Island between 1720 and 1740. This fortress was not simply a military installation; it embodied the culmination of financial resources and engineering acumen, standing as a bulwark against rival empires. The French Royal Treasury dispatched numerous heavily armed pay ships laden with gold, supplies, and imperial secrets across the vast Atlantic to supply such an edifice. Yet the vast, unpredictable ocean—with its tempestuous storms and hidden shoals—proved to be an adversary as formidable as any army. In 1746, one such pay ship—commissioned by the Duc d'Anville on a mission to reclaim Louisbourg—vanished amid a raging tempest. Accounts differ;

some say that in a desperate attempt to protect the treasure, the crew concealed the cargo, while others contend that experienced pirates intercepted the vessel and spirited its bounty away. Whatever the precise events, this maritime calamity added another enticing layer to the Oak Island enigma, forming a distinct theory proposing that lost riches swept away by the Atlantic might be intricately connected to the island's secret history.

The modern chapter of the Oak Island saga began in 1795 when local treasure hunters—moved by generations of whispered legends and the faded remnants of ancient maps—discovered a seemingly unremarkable depression in the landscape. At first, it was dismissed as a natural sinkhole, the result of ordinary geological processes. However, subsequent exploration revealed evidence that this depression was anything but random. Beneath the surface lay a meticulously engineered labyrinth known today as the Money Pit. This subterranean network, consisting of meandering tunnels, timber platforms, and a sophisticated system of flood controls, defied natural explanation. Early excavators reported encountering layers of logs, often spaced at 10-foot intervals, embedded with strange materials like coconut fiber, putty, and charcoal. At 90 feet down, they purportedly found a stone tablet with undeciphered symbols, famously translated by some as: "Forty feet below, two million pounds are buried." This complex construction, complete with alleged booby traps and ingenious flood tunnels seemingly designed to thwart any attempt to reach the bottom, cemented the Money Pit's reputation as a marvel of ancient engineering, fueling the belief that a treasure of immense value lay hidden within. Every stone, every carefully chiseled cross mounted along its walls, appeared to have been deliberately arranged by artisans with a singular purpose: to confound the unworthy and to reward only the dedicated seeker with clues to a hidden legacy.

A turning point in the theories surrounding Oak Island emerged with what modern investigators and proponents of the Templar connection have called the "Impossible Coincidence Theory." This controversial concept posits that Oak Island was designed as a comprehensive, geometric treasure map by purported master architects of the Templar legacy. According to this theory, when harmonized with strategically placed stone monuments, the island's natural contours yield an elaborate code that points, with mathematical precision, to one definitive location. For instance, proponents meticulously analyze the placement and orientation of features like Nolan's Cross (a series of large boulders forming a cross shape on the surface) and argue that its precise angles and measurements, when combined with other landforms, align to reveal specific coordinates. At the heart of this encoded message is the Stone Triangle, an arrangement of massive boulders that appear to serve as a marker or coordinate, pinpointing the locus of the hidden treasure. When considered alongside other enigmatic landmarks—such as the mysterious Nolan's Cross and the long-lost original H-O Stone (the whereabouts of which have been obscured by time)—every deciphered cipher on the island, including the infamous 90-Foot Stone Cipher, seems to converge on one monumental conclusion. For over two and one-third centuries, countless treasure hunters have roamed the island in search of riches within the Money Pit, unaware that the secret lay in the integrated geometry of Oak Island's landscape.

THE SHATTERED CROSS

Far from mere speculation, the "Impossible Coincidence Theory" is presented as a rigorous, multidisciplinary treatise—a challenge to the modern reader to bring together historical insight, precise geometric measurements, and advanced analytical techniques to decipher a hidden language. In this framework, the Money Pit may be understood not as the repository of the ultimate treasure but as a cleverly constructed decoy—one element in a much larger, deliberately designed system intended to mislead all but those who hold the valid conceptual key. This key, some argue, is found not in any single artifact but in the interplay between natural landforms and human ingenuity—a synthesis of ancient geometry, esoteric symbolism, and the refined art of clandestine communication.

The strength of this theory is further bolstered by extensive international research. In Spain, dedicated scholars have scoured ancient cathedrals, monasteries, and archival manuscripts, uncovering Templar inscriptions and cryptographic motifs that proponents argue closely resemble features later identified on Oak Island. Detailed examinations of Spanish religious architecture are cited as evidence for the theory that geometric patterns and symbolic sequences—used to encode secret knowledge—were a widespread and sophisticated practice. Such studies are thus presented to support the idea that the encryption systems developed by the Templars were not isolated innovations but part of a continent-wide method for safeguarding treasured lore.

Portuguese research offers another illuminating perspective. After suppressing the Templar Order in most parts of Europe, many surviving members sought refuge in Portugal, where they transitioned into the Order of Christ under King Denis I. Archival treasures from institutions such as the

Convento de Cristo include elaborate maps, charts, and written records that detail transatlantic navigational routes and, more intriguingly, encode hidden instructions—meant for use during perilous voyages. Among these documents, several 15th-century Portuguese maps contain marginalia and subtle outlines that bear an uncanny resemblance to regions of modern-day Nova Scotia. This observation fuels the theory among certain researchers that Portuguese Templars contributed to a corpus of geographic and cipher techniques that later played a role in forming Oak Island's cryptic layout.

Scotland further supplements this unfolding mystery with its wild highlands and ancient stone circles. Local legends firmly entwined with historical facts indicate that exiled Templars intermingled with native clans, their secret traditions transmitted down through families such as the Sinclairs. The iconic Rosslyn Chapel has long been at the center of these discussions. Its maze of carved figures, hidden recesses, and meticulously calibrated geometrical alignments have led modern scholars to argue that the chapel functions as an elaborate cipher—a visual repository of medieval secrets meant to guide those who can decipher its language. Recent advances in digital imaging and geometrical analysis have uncovered compelling parallels between Rosslyn's cryptic masonry and the enigmatic stone arrangements on Oak Island, leading proponents to suggest that a unified tradition of encrypted knowledge extends from the rugged terrain of Scotland to North America.

Finally, Malta contributes a vital Mediterranean flavor to this international mosaic. Renowned for its impeccably preserved fortifications and its storied legacy as a crossroads for crusading orders, Malta is rich in examples of secret military engineering. Archaeologists and historians working in Malta have documented that many ancient monuments and fortified structures incorporate exacting geometric alignments, deliberate spatial divisions, and cryptic inscriptions that mirror the design principles later observed in Templar constructions. Detailed surveys of Maltese fortifications reveal that the same mathematical precision and symbolic nuance are interpreted as being found in Oak Island's engineered layouts. Such findings are cited as evidence for the theory that the methods of cryptographic architecture developed in the Mediterranean were broadly disseminated—and may have been a critical influence on the creation of the enigmatic treasure map purportedly embedded in Oak Island's very terrain.

Modern technological advancements empower researchers to explore these age-old mysteries with unprecedented clarity. High-resolution satellite imagery, ground-penetrating radar, and sophisticated digital mapping techniques are employed to examine Oak Island's topography in minute detail. These technologies have already begun to reveal previously undetected tunnels, subtle stone alignments, and other anomalies that further support the view of Oak Island as a deliberately engineered map. While no definitive answers have yet been revealed, every new digital scan, every refined measurement, and every archival discovery from repositories in Spain, Portugal, Scotland, or Malta adds another indispensable piece to this intricate puzzle. Together, they form a steadily emerging picture that promises to reveal, piece by piece, the true legacy of the Knights Templar embedded within the island's contours.

Beyond the annals of academia and laboratories of digital archaeology, the Oak Island mystery has exerted a profound cultural impact. Over the decades, storytellers, filmmakers, and writers have transformed the legend into a global phenomenon. The highly popular television series "The Curse of Oak Island" has brought the relentless, decades-long search to millions worldwide, creating a legion of new enthusiasts. Documentaries, books, and television series have broadcast the mystery into millions of homes worldwide, reinforcing that some secrets are so monumental that they defy the boundaries of time and space. Its shared cultural momentum has spawned new generations of researchers and adventurers who see Oak Island as a repository of lost treasure and a living archive. History, myth, and modern science coalesce in an ongoing dialogue on this site.

Of course, not everyone is convinced by the more elaborate theories. Skeptics abound who argue that the "Impossible Coincidence Theory" is little more than wishful thinking, that the coincidences in stone alignments and cryptic symbols can be explained by natural processes or by the tendency of the human mind to find patterns where none were intended. They often point out that countless natural formations can be interpreted as intentional designs if one is looking for them, a phenomenon known as pareidolia, and that historical data is often reinterpreted to fit pre-existing theories. Debates in academic circles continue to rage over-interpreting these supposedly encrypted messages, the statistical likelihood of these "coincidences" arising naturally, and the authenticity of the alleged Templar symbols on the island. Nevertheless, even as controversy persists, the sheer allure of the mystery remains intact. The very debates and controversies fuel renewed research, driving scholarly inquiry and public fascination in a cycle of discovery that appears unlikely to end anytime soon.

Looking to the future, many in archaeology and historical research are optimistic that ongoing technological innovations will one day break through the veil of mystery that has long surrounded Oak Island. New tools—such as artificial intelligence for pattern recognition and advanced three-dimensional modeling—can already reveal previously unnoticed features in archaeological sites. It is conceivable that within the next decade, these innovations will yield definitive evidence that either confirms or refutes the long-held theories regarding the island's origins and the secret legacy of the Knights Templar. For many researchers, finally unlocking this encrypted treasure map represents a triumph over history's mysteries and an opportunity to redefine our understanding of legacy, secrecy, and the human spirit's eternal drive to explore the unknown.

For those who have dedicated their lives to the Oak Island enigma, the journey is as important as the ultimate destination. Each discovery—an overlooked document in a forgotten Spanish monastery, a newly discovered tunnel on the island itself, or a groundbreaking technological breakthrough—brings us incrementally closer to an integrated understanding of the past. Such discoveries can reinvigorate the mystery and stimulate further inquiry, ensuring that the adventure continues unabated despite decades of research.

In the final analysis, Oak Island is far more than a physical repository of lost riches. It has become a monument to the indefatigable human spirit—a living archive of secrets that invites us to explore the very idea of hidden knowledge. The legacy of the Knights Templar, the secretive brilliance of Renaissance thinkers like Bacon, and the daring exploits of maritime adventurers all converge here to form a mosaic of human endeavor that challenges our conventional understanding of treasure. Each deciphered symbol, each meticulously measured alignment, and every archival discovery contributes to a grand narrative that calls on us to rethink the boundaries between myth and reality.

As you turn these pages and join the countless seekers devoted to unraveling the Oak Island mystery, remember that the true treasure may not be measured solely in gold or ancient relics. Instead, it may reside in the profound insights we gain about our cultural heritage, the art of encrypted communication, and the timeless human drive to seek out hidden truth. In every engraving, every ancient marker, and every piece of data uncovered with modern technology, we inch closer to unlocking a legacy that bridges continents and centuries—a legacy that, once revealed, may forever transform our understanding of history, treasure, and the art of discovery.

Thus, the story of Oak Island is far from over. It is an evolving, multilayered narrative challenge cast down by time, inviting each generation to piece together an intricate puzzle that spans the ages. Whether you are a scholar, an adventurer, or simply a curious seeker of truth, your journey into the enigma of Oak Island is a testament to the enduring power of mystery. Every discovery, every analytical breakthrough, and every archival revelation serves as a stepping stone on a path that continues to inspire academic inquiry and popular imagination.

The final key remains elusive in this interplay of stone, water, and ancient code—hidden among the integrated clues of natural formation and human craftsmanship. However, precisely, this elusive quality challenges us to continue our quest, driven by the conviction that the truth will emerge one day. The final key to Oak Island's enigma is not contained in any single artifact; it is written in the language of our collective past—a language whose decipherment promises to unveil material riches and an enduring legacy of secret wisdom.

Chapter Twelve
The Mystery of Rosslyn Chapel

Rosslyn Chapel, situated near Roslin in Midlothian, Scotland, has long been a locus of scholarly inquiry and public fascination. Constructed in the latter half of the fifteenth century by Sir William Sinclair, the 3rd Prince of Orkney and Earl of Caithness, its elaborate Gothic design and intricately carved stone surfaces invite a rethinking of its purpose—not merely as a house of worship but as a potential repository for secret knowledge and esoteric traditions. From its unique architectural elements to its profound symbolic motifs, the chapel has become a crucible for theories suggesting it was designed to encode deeper messages and conceal storied connections with influential medieval organizations.

The construction of Rosslyn Chapel began in 1446, a period of significant cultural and political ferment in Scotland. William Sinclair, a powerful and highly educated nobleman, spared no expense in its creation. Unlike typical parish churches, Rosslyn was built as a collegiate chapel, intended to serve as a place of prayer for a small community of priests and to provide divine services for the Sinclair family. However, the sheer scale and opulence of its design far exceeded the needs of a small collegiate foundation, hinting at a grander ambition or purpose from its inception.

One widely discussed hypothesis involves the influential Sinclair family, whose heraldic symbols and preserved family correspondence imply they may have inherited a custodial role in maintaining aspects of Knights Templar lore. The theory suggests that several knights found refuge in Scotland after the Templar Order was suppressed in 1307. King Robert the Bruce offered them asylum in defiance of the Pope and needed skilled warriors. These Templars, proponents argue, integrated with powerful Scottish families, most notably the Sinclairs (descendants of the French Saint-Clair lineage, from which Templar founders also hailed). This clandestine integration allowed for the unbroken transmission of Templar knowledge and wealth safeguarded through generations within the Sinclair line. Although direct documentary evidence definitively linking the Sinclairs to the Templars is limited and remains a subject of intense debate among mainstream historians, several secondary sources and family traditions point to a deliberate transmission of secretive knowledge, a theory notably advanced by scholars such as Helen Nicholson and popular authors like Christopher Knight and Robert Lomas.

Closely examining Rosslyn Chapel's stonework reveals recurring geometric patterns, spirals, and animal motifs that many art historians contend operate as part of a symbolic language rather than mere decorative ornamentation. Unlike any other medieval building in Britain, the chapel's interior is an explosion of detail. Every surface, from the vaulted ceilings to the column bases, is adorned with carvings. Among the most famous and debated carvings is the "Apprentice Pillar," an intricately twisted column adorned with flowing foliage and figures. Its dramatic legend claims it was carved by an apprentice more skillfully than his master, leading to the master's jealous murder. Beyond its dramatic tale, the pillar's unique design and the presence of numerous "Green Man" motifs—foliate faces representing nature and rebirth, familiar in medieval art but exceptionally abundant here—are often interpreted as alluding to pre-Christian pagan beliefs, esoteric traditions, or even a hidden alchemical significance. Some researchers have interpreted these carvings as instructions or maps revealing the locations of concealed relics and hidden chambers.

Among the more contentious elements is the depiction of maize, or Indian corn, rendered in stone in a context where such imagery is profoundly anachronistic. Given that the chapel predates conclusive European contact with the New World by decades, this motif has sparked debate: some propose that it was used to invoke themes of fertility and pastoral abundance, or perhaps that the carver imagined an exotic plant based on travelers' tales. However, a more provocative theory, popular among those who posit pre-Columbian transatlantic voyages, suggests that the corn motif is direct evidence of contact with the Americas before Columbus. Proponents argue that Templars or their successor organizations, having inherited sophisticated navigational knowledge, embarked on voyages to the New World, bringing back physical evidence and symbolic representations of their discoveries to be secretly encoded within the chapel's fabric.

This specific carving is often cited by proponents of the Templar connection to Oak Island as significant evidence. The theory posits that the same secret society responsible for the complex engineering of Oak Island's Money Pit and its purported geometric treasure map also influenced the symbolic design of Rosslyn Chapel. The presence of the corn motif, therefore, is interpreted as an explicit reference to the North American continent, implying that Rosslyn serves as a complementary 'key' or a symbolic endorsement of the Templar-linked knowledge that extends across the Atlantic to sites like Oak Island. For those who believe in a grand Templar design, the corn at Rosslyn is not merely an artistic anomaly but a deliberate coded message, tying the Scottish chapel directly to the purported pre-Columbian voyages and the hidden secrets of Nova Scotia.

Whatever its original intent, the corn motif contributes significantly to the layered complexity of Rosslyn's iconography and continues to fuel tantalizing speculation.

Local traditions and archival documents add further intrigue. Oral histories preserved among the inhabitants of the surrounding region, combined with Sinclair family manuscripts dating from the early eighteenth century, hint that the chapel may have been engineered with concealed spaces—hidden vaults or crypt-like chambers intended to secure precious relics and encrypted documents during tumultuous times. These legends often suggest hiding significant artifacts, such as the Holy Grail, the Ark of the Covenant, the Templar treasure salvaged from their fall, ancient biblical texts, or even evidence of Templar connections to early Christian history and the lineage of Christ. Modern digital archaeological techniques, including ground-penetrating radar, 3D scanning, and drone-mounted LiDAR, have detected subtle anomalies within the chapel's fabric that some interpret as evidence for such secret passages, particularly beneath the floor of the Lady Chapel and within particular walls. While these non-invasive surveys confirm the presence of subsurface disturbances, definitive archaeological confirmation of hidden chambers containing treasure or ancient secrets remains elusive. Nevertheless, these investigations lend empirical support to longstanding theories that the medieval builders of Rosslyn Chapel intended it to function both as a public monument of piety and as a clandestine container of guarded tradition.

Another provocative dimension to the mystery is the speculative connection with Scotland's national hero, Robert the Bruce. Some alternative historical narratives suggest that Bruce may have found a common cause during his struggle for Scottish independence with exiled Templar remnants. The most famous legend, often recounted in Masonic traditions, claims that a contingent of Templar knights, having escaped persecution, fought alongside Bruce at the pivotal Battle of Bannockburn in 1314. Their sudden appearance and formidable fighting skill are said to have turned the tide against the English, contributing to Scotland's victory and thus cementing a lasting, secretive bond between the Crown of Scotland and the surviving Templars. These accounts argue that his defiance of centralized authority and his reputed alliance with influential families like the Sinclairs could have fostered an environment conducive to preserving Templar knowledge and possibly even their treasure. Although mainstream historians remain cautious about fully endorsing this association, citing a lack of direct contemporary evidence, the symbolic resonance of Bruce's struggles continues to feature in debates concerning the broader cultural impact of Rosslyn Chapel.

Interdisciplinary research has dramatically expanded our understanding of the chapel's mysteries. Comparative examinations of similar contemporaneous structures in Scotland and Europe reveal that while some iconographic themes recur across many medieval monuments, Rosslyn's design is uniquely layered. Scholars have noted its unusual orientation, potentially aligned with astronomical events and sophisticated geometric patterns that some believe could encode mathematical or cosmological knowledge. The chapel's precise proportions and the apparent deliberate 'misplacement' of specific stones are cited by proponents as further evidence of an intentional, symbolic design beyond mere aesthetic choices. Advances in digital imaging and geophysical surveying techniques have confirmed that the building's fabric contains subtle discontinuities that may be deliberate. These modern approaches, detailed in academic publications such as those in the Journal of Archaeological Science, illustrate how the interplay of documented evidence and digital methodologies reshapes our comprehension of medieval art and architecture.

Moreover, contextualizing Rosslyn Chapel within its broader socio-political and religious milieu of late-fifteenth-century Scotland offers additional insight. It was an era marked by resistance to centralized ecclesiastical and secular power, particularly as the Scottish kingdom navigated its precarious relationship with the English crown and the Papacy. The looming shadow of the Protestant Reformation in Europe, and eventually in Scotland, created an atmosphere where the desire to preserve traditional or new forms of knowledge often took on a clandestine character. The possibility that the chapel was intended to serve a dual function—as both a public monument and a secret archive—reflects the complexities of parliament, piety, and power during turbulent times. In such a setting, the interplay between overt symbolism and hidden meaning becomes a powerful metaphor for the struggle to safeguard cultural traditions in the face of political and religious change.

Beyond the Templar connection, Rosslyn Chapel holds significant appeal for Freemasons. Many Masonic lodges and researchers see the chapel as a powerful symbol of their traditions and lineage, tracing connections through the stonemasons' guilds of the medieval era. They interpret many of the carvings—including the Apprentice Pillar and various geometric patterns—as containing Masonic symbolism, suggesting that the chapel's builders were either early Freemasons or that the knowledge embedded within it laid the groundwork for Masonic philosophy. While the precise historical link between medieval operative stonemasons and speculative Freemasonry remains debated, Rosslyn Chapel is undoubtedly revered as a spiritual home and a place of deep symbolic resonance within Masonic circles worldwide, further adding to its layers of mystery.

Chapter Thirteen
The Hospitallers & The Fate of Templar Assets

The sudden suppression of the Knights Templar in 1307 sent shockwaves throughout medieval Christendom. An order famed not only for its valor on the battlefield but also for pioneering early financial practices, the Templars controlled vast estates across Europe while operating an innovative network of banking that facilitated trade and secured the passage of pilgrims to the Holy Land. When royal pressure and papal decree abruptly dissolved the order, a complex legacy of wealth, infrastructure, and mystery was left in its wake. The papal decree Ad Providam of 1312 formalized the transfer of many Templar assets to the Knights Hospitaller, setting the stage for one of history's most remarkable successions of power, symbolism, and cultural memory. This papal decision was not arbitrary; Pope Clement V, pressured by Philip IV, sought to prevent Templar assets from falling entirely into secular hands. By assigning them to the Hospitallers, another prominent military order with a long history in the Holy Land and a mission aligned with Christendom, the Church aimed to ensure the resources continued to serve a crusading purpose.

Nevertheless, this transformation was neither smooth nor universally accepted. In kingdoms like France, England, and across the Iberian Peninsula, ambitious rulers—eager to replenish depleted treasuries—wrested control of many Templar holdings, leading to heated legal contests and intense regional rivalries. King Philip IV of France, who orchestrated the Templar downfall, was reluctant to relinquish the immense wealth he had seized, and substantial portions of French Templar property ultimately remained in royal hands. Similarly, King Edward II initially resisted the transfer in England, leading to protracted disputes over the estates. Even as these disputes smoldered, the Hospitallers inherited piles of precious lands and buildings and the very infrastructure of sophisticated medieval commerce. They repurposed these tangible assets into fortified hospitals, naval bases, and administrative outposts that would support their evolving twin missions of military defense and charitable care for centuries. While consolidating power, this economic realignment also laid the fertile groundwork for legendary tales to arise, a gateway through which matter and myth intertwine.

The economic impact was one of the most significant markers of this transition. The Templars' legacy included an intricate system of financial networks—issuing letters of credit, managing deposits, and facilitating international trade—that had become indispensable to medieval Europe's political and commercial fabric. When the order collapsed, the meticulously maintained ledgers and prized properties fell into new hands, triggering a ripple effect that unsettled centuries-old economic practices. The Hospitallers, less steeped in commerce than their Templar predecessors, chose instead to channel the recovered wealth into bolstering physical security and enhancing military readiness. They built impregnable fortresses and established naval outposts, redirecting lost financial innovation into institutional stability. This shift was not merely one of numbers and assets; the Hospitallers' primary focus had always been providing medical care for pilgrims and Crusaders, a mission requiring secure hospitals and supply lines supported by military defense. While they were formidable warriors, their economic activities were more geared towards sustaining their charitable and military operations than engaging in large-scale banking. By acquiring the Templars' vast landholdings and income streams, the Hospitallers saw an opportunity to significantly fortify their presence in the Mediterranean, particularly against the rising power of the Ottoman Empire. Indeed, the Hospitallers were already a well-established and formidable force, renowned for their hospitals and maritime strength, and this acquisition only served to bolster an already robust foundation. This economic realignment set the stage for shifting the narrative from sheer economic might to the fertile ground where mystical legends could unfold.

Parallel to these tangible economic shifts, a rich narrative of mysticism and myth was born. The sudden and brutal disappearance of the Templars only deepened the enigma surrounding them. Legends sprang up regarding secret relics and esoteric wisdom hidden away in the chaos of their final moments. Tales of sacred objects—the Holy Grail, the Ark of the Covenant, cryptic scrolls of ancient knowledge—quickly mingled with the all-too-real loss of treasure and power. As the Hospitallers assumed control over aged fortresses and hidden crypts once held by the Templars, whispers circulated that secret documents or ritual objects had been expertly concealed within these walls. While historical records rarely support these claims with verifiable evidence, the very nature of the Templar mystery encouraged such speculation. Such rumors found fertile ground in Malta, where the labyrinth of underground chambers and imposing bastions was the ideal backdrop for enchanting legends. Local villagers sometimes recalled furtive glimpses of "knights in faded red" disappearing at dusk—a spectral echo of a vanished order whose mystique transcended time. Although the Hospitallers concentrated primarily on pragmatic military strategy and administrative governance, occasional discoveries of enigmatic engravings and unexpected architectural motifs within former Templar properties continue to fan the flames of speculation regarding a clandestine transmission of Templar mysteries. Thus, the interplay between myth and practicality bridges the tangible legacy of wealth with the intangible power of narrative, reinforcing the orders' lasting allure.

The evolution of symbolic identity that accompanied power transfer was no less significant than these legends. The Templars were immediately recognizable by their stark red cross—a bold Pattée emblem worn on plain white mantles that signified sacrifice, piety, and unyielding martial valor. This simple yet powerful symbol resonated with the fervor of the Crusades and epitomized an unwavering commitment to defending Christendom. In contrast, as the Hospitallers absorbed the legacy of the Templars, they forged a new identity by transforming this emblem into the intricate eight-pointed Maltese Cross. Each point of the modern cross is imbued with virtues such as truth, faith, humility, and justice, perfectly reflecting the Hospitallers' broadened mission to fuse martial prowess with humanitarian outreach. In many Maltese fortresses and chapels, decorative motifs reveal subtle echoes of the older Templar inscriptions, underscoring a continuity that connects traditions with reinterpreted modern values. The evolution from the plain red Templar cross to the ornate Maltese Cross is not merely a design change but a powerful metaphor for transformation—a melding of steadfast military valor with the ideals of charity and compassion that simultaneously reassured allies and challenged adversaries.

Malta, in particular, emerged as a vital nexus in this transformation—an island where Templar and Hospitaller legacies converged in intriguing ways. Although formal Hospitaller rule over Malta began in the early 16th century when Emperor Charles V granted the island to the order, evidence suggests that the Templars may have briefly touched Maltese shores long before this period. Malta's strategic location in the heart of the Mediterranean, with natural deep harbors and a key position along pilgrimage routes, made it an ideal refuge for crusaders and a critical staging post for securing dangerous maritime passages. While definitive archaeological proof of a significant, long-term Templar presence on Malta before 1530 remains limited, some historical fragments, faded inscriptions, and resilient local legends speak of transient Templar detachments—those "knights in faded red" whose ephemeral presence left an indelible mark on the island's collective memory. These theories propose that Templars may have used Malta as a temporary base or a transit point during their voyages across the Mediterranean. This early encounter not only prefigured Malta's later fortification by the Hospitallers but also laid a cultural foundation that the Hospitallers would build upon, blending inherited infrastructure with renewed strategic foresight. The Hospitallers later transformed Malta into an impregnable fortress, culminating in their heroic defense during the Great Siege of 1565 against the Ottoman Empire, a testament to their military and engineering prowess.

The reach of these knightly orders was not confined solely to Malta; their influence spread across continents via diverse routes that reflected both documented conquests and tantalizing speculations. In the British Isles, for example, the Templars and the Hospitallers established commanderies that served as centers of administration, defense, and spiritual solace. In Wales and England, fortified preceptories and round churches—models inspired by the Church of the Holy Sepulcher in Jerusalem—became not only vital protective outposts along key pilgrimage routes but also symbols of continuity in a troubled land. Further south, in Italy and Spain, extensive networks of estates and fortresses entrenched the orders within local power structures. These regions witnessed the fusion of local customs with the external prestige of the Templar legacy, whether the holdings remained independent bastions or were gradually absorbed by the emerging feudal order. In Portugal, a dramatic twist unfolded when King Dinis, defying the Pope's decree, reformed the remnants of the Templar order into the Order of Christ. It was a crucial move that allowed Templar assets to remain largely within Portugal's control rather than being fully transferred to the Hospitallers. Melding ancient martial acumen with groundbreaking navigational skills, the Order of Christ proved vital, as it would later fuel Portugal's Age of Exploration. Their distinctive red cross, often bearing a white fimbriation (an outline), proudly adorned the sails of Portuguese caravels that embarked on voyages across unknown seas, establishing contact with distant lands and influencing colonial endeavors in Latin America through architectural resemblances, ritual traditions, and iconographic motifs. It is in this latter context that New World connections emerge. While medieval documentation of a direct Templar or Hospitaller presence in the Americas before Columbus is scant, several speculative theories have captivated modern enthusiasts. For instance, narratives persist with minimal verifiable evidence that a fragmentary Templar presence might have extended to Nova Scotia, where maritime routes and local legends hint at possible rendezvous points for fugitive knights. The famous Oak Island mystery, with its cryptic clues and tales of buried treasures, is sometimes woven into the fabric of these theories, suggesting that secret caches left by the Templars or their Hospitaller successors may still lie hidden beneath New World soils. Proponents of these theories often point to alleged stone carvings, unique architectural anomalies, or reinterpreted indigenous oral traditions as tantalizing, albeit unproven, clues. Modern successors of these orders, such as the Sovereign Military Order of Malta (the modern incarnation of the Hospitallers), maintain a robust international presence and continue to engage in humanitarian efforts across the Americas—further extending the symbolic legacy of their medieval forebears into the modern New World. While these New World connections remain partly speculative and steeped in legend, they serve as a reminder that the reach of these medieval orders transcended conventional boundaries, inspiring countless myths and intriguing historical debates that persist to this day.

Ultimately, the remarkable journey from the Templars' dramatic dissolution to the Hospitallers' enduring reinvention encapsulates a profound reimagining of medieval Europe's military, economic, and spiritual identity. The Hospitallers not only absorbed a vast legacy of wealth, administrative prowess, and symbolic heritage inherited from the Templars but also transformed these elements to meet their time's shifting geopolitical and cultural realities. Their impressive longevity and independent achievements, such as their heroic defense of Rhodes and Malta, testify to their distinct identity and mission. The evolution from the plain red

Templar cross to the ornate, multifaceted Maltese Cross is a compelling visual metaphor for this transformation—a melding of martial valor with a renewed dedication to humanitarian service. However, for an informed reader, it is crucial to temper the allure of myth with historical pragmatism. While the Hospitallers certainly benefited from the Templars' immense wealth and infrastructure, mainstream historical scholarship emphasizes that the transfer of assets was essentially a practical, politically driven act, not an esoteric inheritance. The Hospitallers were a distinct order with established rules, missions, and identity. There is little to no credible historical evidence to suggest that they consciously perpetuated Templar secret rituals, hidden knowledge, or specific Templar esoteric traditions. Their primary focus was maintaining their charitable work and defending Christendom, particularly in the face of the Ottoman threat. The "secret caches" and "knights in faded red" are essentially part of popular folklore and later romantic interpretations rather than verifiable historical events. The enduring mystique surrounding this power transfer largely stems from the dramatic nature of the Templars' fall and the subsequent void left in the public imagination, which popular narratives eagerly filled. Whether encountered in the shattered remains of medieval fortresses, the rolling landscapes of the British countryside, or the whispered legends of far-flung frontiers, the echoes of these intertwined orders continue to captivate the modern imagination. In this richly woven tapestry of movement, memory, and myth, the legacy of the Templars and the Hospitallers endures as a testament to the transformative power of chivalric ideals, the resilience of medieval institutions, and the ever-adaptive spirit of those who strive to serve a higher cause.

However, it is crucial to acknowledge the skeptical perspective that grounds the discussion in mainstream historical and archaeological consensus. Many academic historians view Rosslyn Chapel primarily as a magnificent example of late Gothic architecture, reflecting the immense wealth, piety, and artistic taste of its patron, William Sinclair. They argue that the "mysteries" often result from modern misinterpretations, pareidolia (the human tendency to find patterns in random data), or the romanticization of medieval symbolism. For example, corn's presence is sometimes explained as a general symbolic representation of abundance rather than a literal depiction or as a later addition by a restorer familiar with New World flora. Furthermore, they emphasize that while the Sinclair family was powerful and connected to figures from the Crusades, direct, verifiable evidence linking them to the Templars' secrets or treasures is conspicuously absent from contemporary historical records. From this viewpoint, Rosslyn Chapel's true wonder lies not in hidden secrets but in its unparalleled artistry and the rich tapestry of its *documented* history, compelling enough without the need for unproven theories.

The chapel's fame surged dramatically in the early 21st century, particularly due to its prominent role in Dan Brown's hugely popular novel, "The Da Vinci Code" (2003), and its subsequent film adaptation. Notably, many of the core speculative theories linking Rosslyn Chapel to the Knights Templar, the Holy Grail, and a secret lineage were significantly popularized by the 1982 non-fiction book *The Holy Blood and the Holy Grail*. Brown's narrative explicitly links Rosslyn Chapel to the Knights Templar, the Holy Grail, and the secret lineage of Jesus Christ, propelling it into the global spotlight and attracting unprecedented numbers of visitors. This surge in popular culture has significantly amplified its mystique, transforming it from a regional architectural gem into an internationally recognized symbol of ancient secrets and hidden truths. While the book is a work of fiction, its influence on public perception of Rosslyn's enigmatic qualities is undeniable, cementing its status as a site of enduring mystery.

Rosslyn Chapel represents a unique intersection of documented medieval artistry, secretive tradition, and modern scientific inquiry. Its richly nuanced carvings, the provocative possibility of hidden chambers, and the alleged custodianship of Templar lore by the Sinclair family—supplemented by even tentative connections to figures like Robert the Bruce and its strong ties to Freemasonry—create a vibrant tapestry that challenges simple interpretation. While many of these theories remain subjects of ongoing scholarly debate, the cumulative evidence supports an interpretation of Rosslyn Chapel as a dynamic monument where truth and myth converge, reflecting an enduring human drive to reveal and obscure the sacred. Whether it holds physical treasure or ancient secrets, Rosslyn Chapel undeniably remains a profound testament to the enduring power of symbolism and the human imagination.

Chapter Fourteen
Templar Influence on Secret Societies & Freemasonry

For centuries, the Knights Templar's mysterious legacy has captivated scholars and the public alike. Even after their dramatic suppression in 1307—when their vast wealth, secret rituals, and purported treasures appeared to vanish from official records—the mythic aura of the Templars only grew stronger. This enduring legend eventually found fertile ground among secret societies and Freemasonry, with popular narratives of secret continuities becoming entwined with medieval memory during the Enlightenment and Romantic eras. During these periods, writers and thinkers transformed elusive historical fragments into a resplendent tapestry of symbolism and ritual, crafting a narrative that suggested a hidden stream of Templar wisdom persisted. In this chapter, we explore the historical claims linking Templar traditions to early Freemasonry, examine how modern secret societies have embraced and reinterpreted Templar imagery, and critically assess whether any authentic Templar rituals persist today, acknowledging the profound cultural and psychological impact of these enduring legends.

In the immediate aftermath of the Templar dissolution, Europe was rife with rumors, intrigue, and political opportunism. With their immense fortunes disappearing overnight, contemporaries speculated about the fate of their material assets and the survival of the order's sacred knowledge. Early polemical tracts and clandestine manuscripts, often written under secrecy, hinted that secret rituals and symbols had been hidden in plain sight—perhaps within the stonemason guilds that later evolved into the lodges of modern Freemasonry. One striking phrase from a surviving medieval document, often cited by proponents of this theory, declares, "...as the light of the holy order was extinguished in the open, its wisdom shone secretly in the halls of the makers of stone." Such evocative lines captured the imagination of later generations, spurring many to believe that the Templars, forced underground by persecution, became the unsung progenitors of secret rites, silently continuing their esoteric traditions and guarding profound, possibly forbidden, knowledge that encompassed Gnosticism, Kabbalah, or ancient wisdom about sacred geometry.

The intellectual ferment of the Enlightenment further transformed these early legends. Writers and scholars—fascinated by the mystery of lost chivalric orders—began explicitly establishing connections between the Templars and the emerging lodges of Freemasonry. A pivotal moment came with Chevalier Andrew Michael Ramsay's 1737 Oration, delivered to a Masonic lodge in Paris, which eloquently linked Freemasonry to the Crusader orders and a chivalric, Christian lineage, paving the way for Templar-themed degrees. Later in the 18th century, Baron Karl Gotthelf von Hund's Strict Observance Rite in Germany explicitly claimed a direct, secret descent from the Templars, complete with grand masters supposedly holding a hidden chain of command. In the late 20th century, Stephen Knight observed, "The mythic narrative of the Templars serves as a repository for the aspirations of secret societies, irrespective of any direct historical lineage." In similarly influential works, Peter Partner and Alain Demurger have argued that modern Masonic ritual is best understood as an allegorical construction, drawing upon a rich reservoir of medieval symbols and narratives rather than an unbroken line of specific rites. Such scholarly counterpoints underscore that while the Templar tradition is not documented as having passed intact into later societies, it nonetheless provided a fertile archetype for constructing secret symbolism. This period saw a convergence of interest in ancient wisdom, chivalric ideals, and the allure of secrecy, making the Templar narrative a perfect fit for the burgeoning esoteric movements.

A vivid timeline helps to situate these developments and trace the legend's evolution. From 1200 to 1307, the Knights Templar existed as a cohesive, militarized order characterized by disciplined rituals, strict hierarchies, and the distinctive red Pattée cross—a symbol of sacrifice, piety, and unwavering commitment. Their grandeur is recorded in illuminated manuscripts and architectural monuments, evidencing spiritual zeal and practical prowess. Then, in the turbulent period following their suppression from 1307 to 1500, scattered accounts, oral traditions, and secretive texts alluded to the fate of Templar lore. Amid political instability, surviving members may have sought refuge in underground networks or allied themselves with local craft guilds, which prized secrecy and skilled artistry, particularly in regions like Scotland and Portugal where the order's persecution was less severe or where its members found new guises, such as in King Dinis's Order of Christ. Candlelit manuscripts and whispered exchanges gradually transformed these whispers into enduring legends. From 1700 to the present, modern Freemasonry and similar secret societies emerged as distinct entities. During this period, many Masonic rituals and symbolic systems were assembled and elaborated upon, drawing on available historical fragments and the imaginative energies of the Enlightenment and Romantic movements. Today's secret rituals—replete with allegorical narratives of lost temples and cryptic keys to divine wisdom—reflect a modern reinvention of what once belonged to the Templars, at least in legend. This reconstructed timeline is not merely a framework for historical debate but a narrative journey connecting illuminated medieval manuscripts with gilded Masonic regalia. Many lodges feature the Red Cross and other geometric emblems alongside linguistic allusions to ancient temples and hidden vaults. Such symbolic layering offers modern members a sense of continuity with a storied past, even if the "lineage" is best understood as a mythic echo rather than a direct transmission of ritual.

Modern Masonic lodges and neo-Templar groups adopt numerous symbols that they claim bear the stamp of ancient tradition. The Templars' stark red cross remains a powerful image that modern secret societies often juxtapose with symbols such as the square and compasses, the all-seeing eye, and various depictions of sacred geometry. These images frequently appear in ritual regalia, ornamental carvings, and on the walls of meeting halls, functioning as bridges between an idealized medieval past and contemporary aspirational ideals. In many lodges, ceremonies invoke allegorical narratives of a "lost temple" or a "secret vault" containing keys to eternal wisdom. For instance, the Knight Templar degree within the

York Rite of Freemasonry explicitly reenacts elements of Templar history and symbolism, albeit in a purely allegorical context focusing on moral lessons. These ritual elements are sometimes accompanied by reciting passages drawn from reinterpreted medieval texts, reinforcing that a hidden tradition persists beneath the surface of formal history.

While the interplay of history and myth is captivating, the need to separate fact from fiction is equally critical. Despite the powerful visual and ritual connections, unambiguous evidence of an unbroken transmission of Templar practices into modern Freemasonry is notoriously scarce. Some neo-Templar advocates claim that secret rites have been preserved through generations and passed down in private ceremonies, often citing supposed clandestine gatherings in Scotland or other sympathetic regions after the central suppression. However, a preponderance of scholarly opinion—supported by careful exegesis of both medieval documents and modern ritual manuals—suggests that what is now seen as "Templar influence" is essentially the product of later creative reinterpretation rather than a direct survival of 13th-century practices. One modern historian argues, "The Templar influence in secret society is best conceptualized as an echo—a beautifully refracted memory that has been transformed by time and rearticulated to suit modern sociocultural needs." This distinction is vital because it underscores that modern secret societies have adopted Templar symbols not through an unbroken chain of heritage but through a process in which ancient images are reimagined to evoke ideals such as honor, sacrifice, and eternal vigilance. Far from being static relics, the symbols of the Templars have become dynamic components of modern ritual culture. They are used as visual shorthand for values that resonate with contemporary seekers of esoteric knowledge, even as the precise details of medieval Templar rites remain lost mainly.

Beyond the ritual halls of Freemasonry, the Templar legacy occupies a significant space in popular culture. From literature to cinema, the image of the Templar knight shrouded in mystery has been repeatedly reinterpreted. Films like *National Treasure* and novels such as *The Da Vinci Code* have popularized the idea that the Templars guarded secret documents and hidden treasures, ideas that, in turn, continue to influence the internal iconography and rituals of modern secret societies. This cultural penetration has reinforced popular expectations and, in many cases, provided additional themes for Masonic and neo-Templar groups to incorporate into their mythologies. It is also worth noting that the influence of these symbols is not confined solely to ritual. In specific political and social contexts, Templar imagery has been invoked as a metaphor for resistance against perceived oppression. It is evident in various modern movements where the idea of an underground, secret legacy of righteous warriors is employed to mobilize purpose and create a sense of shared identity. Whether in political manifestos, in artistic installations, or the very design of public monuments, the enduring allure of Templar symbolism speaks to a broader cultural resonance that underscores the complex relationship between myth and collective memory.

At the heart of all these developments lies a vibrant scholarly debate. On one side stand those who maintain that a genuine, if altered, continuity exists between medieval Templar practices and modern secret societies, arguing that Templars, through careful planning, passed their knowledge to select successors. On the other side are scholars who contend that the

Templar legacy primarily exists as an imaginative reconstruction—a narrative drawn upon to supply modern rituals with an air of ancient authority rather than reflecting direct transmission. The debate is animated by contrasting voices: neo-Templar proponents who insist on preserving the age-old mystery and critical historians who argue that modern secret rituals are the creative products of more recent cultural forces influenced by Masonic thought and the Romantic fascination with the medieval past. Often, the division is expressed through the language of authenticity versus reinvention. As one prominent academic once summarized, "Modern secret societies do not so much resurrect Templar ritual as they reinvent it in the light of contemporary aspirations for mystique and moral rectitude."

In conclusion, the influence of the Knights Templar on secret societies and Freemasonry is a layered and manifold phenomenon. Early legends regarding the Templar refuge in stonemason guilds set in motion symbolic transmission, whereby ancient emblems and ritual language were reinterpreted over centuries. Building on these mythic foundations, modern secret societies have adopted and adapted Templar symbols—such as the Red Cross, sacred geometry, and allegories of lost temples—as potent metaphors for values like honor, courage, and the quest for secret knowledge. While the historical record does not support an unbroken line of ritual transmission from the medieval Templars to contemporary Freemasonic practice, the power of myth has ensured that their archetypal images continue to resonate. In this interplay between complex history and imaginative mythmaking, the Templar legacy endures—as both inspiration and enigma, a call to the mysterious, the secret, and the eternal search for hidden truth. As we reflect on these themes, it is worth asking: Are the rituals we witness today authentic relics of an ancient past, or are they modern constructs re-envisioned to meet our contemporary yearning for mystery and connection? Perhaps the answer lies not in a binary of truth versus invention but in appreciating the Templar legacy as an evolving dialogue that connects medieval valor with modern aspirations for wisdom, community, and Enlightenment.

Chapter Fifteen
The Vatican Archives & Missing Templar Documents

Few mysteries in medieval history capture the imagination quite like the fate of the Knights Templar. In 1307, their abrupt suppression led not only to the seizure of their immense wealth and privileges but also, it is widely believed, to a deliberate effort to erase—or secretly preserve—their records. Despite official narratives labeling the order as heretical, persistent rumors have maintained that the true story of the Templars is hidden within the shadowy vaults of the Vatican Archives. Central to these intrigues is the Chinon Parchment, a document drafted in 1308 at the Château de Chinon, which famously reveals that Pope Clement V, acting in secret, had privately absolved many Templars of heresy, clearing them of the gravest charges, even as he publicly dissolved their order under political duress. An evocative line from an anonymous medieval document resonates with this idea: "...as the light of the holy order was extinguished in the open, its wisdom shone secretly in the halls of the makers of stone." This language, imbued with a sense of clandestine preservation, hints at an intentional strategy by ecclesiastical authorities to safeguard aspects of Templar wisdom—acknowledging, as scholars like Stephen Knight suggest, that the Templars' contributions both in martial prowess and esoteric knowledge may have been too valuable to vanish entirely.

In the immediate aftermath of the Templars' downfall, Europe became rife with rumors, intrigue, and political maneuvering. Political forces and ecclesiastical interests converged as many contemporaries speculated that the Templars had transferred their most sensitive records into hidden forms, secret correspondences, and ritual manuals, documents deemed too dangerous or politically disruptive for public knowledge. Consequently, the Vatican Archives, with their historical reputation for restricted access and guarded secrecy, have long been a magnet for those seeking the lost truths of this enigmatic order. It is crucial to note that the "Secret" in "Vatican Secret Archives" (Archivum Secretum Vaticanum) historically meant "private" or "personal" to the Pope, not "hidden" or "clandestine" in a conspiratorial sense. Moreover, despite their stringent rules, the archives are accessible to qualified scholars, albeit under strict conditions, allowing dedicated researchers to delve into their vast holdings. Nevertheless, the archives' controlled access and vast, often unindexed, collections have naturally fueled public fascination and speculation. Beyond the Chinon Parchment itself—which was remarkably rediscovered in the Vatican Secret Archives in 2001 by historian Barbara Frale and made publicly available, confirming the Pope's private absolution—many researchers now argue that the Archives may harbor additional papal records that not only clear the Templars of heresy but also chronicle their secret missions and inventories of relics. Indeed, the Archives contain the extensive Processus contra Templarios, the official trial records of the Templars, which include detailed interrogations and confessions—albeit many extracted under torture—providing a controversial yet invaluable primary source on the order's final days. Rumors abound of detailed ledger-like inventories explaining the disposition of relics—perhaps even fragments of the True Cross—along with encrypted texts detailing covert correspondences between Templar commanders and high-ranking Church officials. Such documents, if ever unearthed from the Vatican's vast collection, which includes millennia of papal bulls, diplomatic correspondence, financial accounts, and various judicial proceedings, could reveal an astonishing narrative in which the Church both condemned and secretly preserved the legacy of an order that once wielded tremendous power. It is also important to remember that many Templar records were openly seized, destroyed, or transferred to other entities, particularly the Knights Hospitaller, by various European monarchs following the order's suppression, meaning the Vatican's holdings represent only one part of the Templars' dispersed documentary legacy.

A particularly captivating element of this ongoing mystery is the notion of a "second archive" within the Vatican—a hidden repository believed by some to be reserved for documents considered too explosive or controversial for mainstream consumption. Although concrete evidence for the existence of such a repository remains highly speculative, its alleged presence has fueled much fascination. Some researchers maintain that this clandestine vault may contain comprehensive records of Templar internal rituals, doctrinal debates, and logistical details that provided the scaffolding for their military and spiritual endeavors. The very idea of a secondary archive encourages the imagination to wander through shadowy corridors where centuries of suppressed history lie dormant, awaiting the probing light of modern research. This, coupled with the Vatican's storied history of controlled document release—often motivated by diplomatic protocols, theological sensitivities, or the sheer volume and complex indexing of its collections—enhances the mystery and suggests that such a repository could be key to understanding not only the Templars but also the broader dynamics between the Church and dissenting orders, or even details of events the Church preferred to keep from public scrutiny.

A vivid timeline further clarifies the evolution of these documents and the legends surrounding them. Between 1200 and 1307, the Knights Templar operated as a disciplined, unified military order, meticulously recording their deeds in illuminated manuscripts and internal administrative documents, often adorned with their emblematic red Pattée cross. In the turbulent period from 1307 to 1500, however, fragmented accounts, oral histories, and clandestine texts began to circulate. During this time, sympathetic church officials, faced with intense political pressure, may have covertly preserved key aspects of Templar heritage, either by reclassifying documents or by ensuring they ended up in less accessible sections of burgeoning archives. From 1700 to the present, the mythologized symbols of these records were reimagined and adopted by emerging secret societies such as the Freemasons, fueling modern legends. Although modern ritual practices are clearly influenced by later reinterpretations rather than direct historical continuity, they nonetheless serve as a living testament to the enduring appeal of the Templar archetype, continually reshaped by cultural, intellectual, and technological developments that seek to find profound meanings in hidden histories.

Contemporary advances in digital archiving and multispectral imaging now offer an unprecedented opportunity to unlock these hidden treasures. Cutting-edge "digital exhumation" techniques allow scholars to reveal once-faded ink, decipher complex ciphers, and reintegrate fragmented documents into coherent texts. Such technological breakthroughs hold the promise that additional Templar records—be they secret correspondences, inventories of relics, or internal ritual texts—may soon be unveiled. The prospect of finally decoding these materials not only excites academic circles but also challenges long-held historical interpretations, potentially leading to a radical rethinking of both the Templars' role in medieval Christendom and the interplay between Church authority and secret orders. Furthermore, these technologies can help in identifying documents that may have been intentionally or accidentally *lost* over centuries due to war, fire, or simple disorganization, adding another dimension to the quest for complete historical truth.

A comparative perspective further enriches this narrative. Whereas institutions such as the British Library or the Russian State Archive have adopted more transparent archival practices in recent years, gradually making vast collections accessible online, the Vatican's historic reputation for stringent control over its collections amplifies the sense that some documents may have been intentionally concealed or are simply extraordinarily difficult to access due to their sheer volume and complex cataloging systems. This guarded stewardship has, in turn, spurred modern conspiracy theories and academic debates that underscore the tension between the desire for public knowledge and ecclesiastical control. As more scholars call for comprehensive digitization projects and collaborative research initiatives, the possibility of long-hidden documents emerging from the Vatican Archives grows ever more tangible. Each new technological advance brings us closer to deciphering the full extent of the Church's secret records, promising to illuminate the dark corners of a history that many believed was irrevocably lost.

Reflecting on these intricacies invokes a series of enduring questions: What hidden treasures, if any, remain in the Vatican Archives that might radically alter our understanding of the Templars? Could the discovery of lost inventories, secret correspondence, and ritual texts fundamentally reshape our view of papal policy and medieval military orders? And, perhaps most provocatively, how will such revelations influence our modern conceptions of power, secrecy, and the transmission of knowledge? The quest to answer these questions is not merely an academic exercise—it stands as a testament to the perennial human desire to reclaim lost truths and to challenge the narratives imposed by powerful institutions. In grappling with these mysteries, we confront not only the shadowy legacy of the Templars but also the broader dynamics of concealment and revelation that have defined the relationship between religious authority and historical memory for centuries.

In conclusion, the Vatican Archives serve not as static repositories of outdated records but as dynamic vaults where the past converses with the present. The Chinon Parchment's quiet act of absolution, made public after centuries, is emblematic of a broader, deliberate strategy by the Church to manage—and perhaps even preserve—the Templar legacy, albeit within controlled parameters. Coupled with persistent rumors of hidden inventories, internal ritual texts, and a possible "second archive," these mysteries challenge us to reconsider the boundaries of documented history and inspire the quest for further discovery. While mainstream historical consensus holds that most Templar records were either destroyed, transferred, or publicly absorbed by other orders, the allure of the unknown within the Vatican's vast repositories continues to fuel speculation. As digital innovations and interdisciplinary research continue to push the limits of archival inquiry, the promise of unveiling the lost chapters of the Templar saga shines ever brighter. Ultimately, the enduring allure of the Templars lies in their capacity to provoke questions that transcend time, urging each new generation to engage with the timeless interplay of power, piety, and secrecy.larly captivating element of this ongoing mystery is the notion of a "second archive" within the Vatican—a hidden repository believed by some to be reserved for documents considered too explosive or controversial for mainstream consumption. Although concrete evidence for such a repository remains highly speculative, its alleged presence has fueled much fascination. Some researchers maintain that this clandestine vault may contain comprehensive records of Templar internal rituals, doctrinal debates, and logistical details that provided the scaffolding for their military and spiritual endeavors. The idea of a secondary archive encourages the imagination to wander through shadowy corridors where centuries of suppressed history lie dormant, awaiting the probing light of modern research. This, coupled with the Vatican's storied history of controlled document release—often motivated by diplomatic protocols, theological sensitivities, or the sheer volume and complex indexing of its collections—enhances the mystery and suggests that such a repository could be key to understanding not only the Templars but also the broader dynamics between the Church and dissenting orders, or even details of events the Church preferred to keep from public scrutiny.

A vivid timeline further clarifies the evolution of these documents and the legends surrounding them. Between 1200 and 1307, the Knights Templar operated as a disciplined, unified military order, meticulously recording their deeds in illuminated manuscripts and internal administrative documents, often adorned with their emblematic red Pattée cross. In the turbulent period from 1307 to 1500, however, fragmented accounts, oral histories, and clandestine texts began circulating. During this time, sympathetic church officials, faced with intense political pressure, may have covertly preserved key aspects of Templar heritage by reclassifying documents or by ensuring they ended up in less accessible sections of burgeoning archives. From 1700 to the present, the mythologized symbols of these records were reimagined and adopted by emerging secret societies such as the Freemasons, fueling modern legends. Although modern ritual practices are influenced by later reinterpretations rather than direct historical continuity, they serve as a living testament to the enduring appeal of the Templar archetype, continually reshaped by cultural, intellectual, and technological developments that seek to find profound meanings in hidden histories.

Contemporary advances in digital archiving and multispectral imaging offer an unprecedented opportunity to unlock these hidden treasures. Cutting-edge "digital exhumation" techniques allow scholars to reveal once-faded ink, decipher complex ciphers, and reintegrate fragmented documents into coherent texts. Such technological breakthroughs promise that additional Templar records—secret correspondences, relic inventories, or internal ritual texts—may soon be unveiled. The prospect of finally decoding these materials excites academic circles. It challenges long-held historical interpretations, potentially leading to a radical rethinking of the Templars' role in medieval Christendom and the interplay between Church authority and secret orders. Furthermore, these technologies can help identify documents that may have been intentionally or accidentally *lost* over centuries due to war, fire, or simple disorganization, adding another dimension to the quest for complete historical truth.

A comparative perspective further enriches this narrative. Whereas institutions such as the British Library or the Russian State Archive have adopted more transparent archival practices in recent years, gradually making vast collections accessible online, the Vatican's historic reputation for stringent control over its collections amplifies the sense that some documents may have been intentionally concealed or are simply extraordinarily difficult to access due to their sheer volume and complex cataloging systems. This guarded stewardship has, in turn, spurred modern conspiracy theories and academic debates that underscore the tension between the desire for public knowledge and ecclesiastical control. As more scholars call for comprehensive digitization projects and collaborative research initiatives, the possibility of long-hidden documents emerging from the Vatican Archives grows more tangible. Each new technological advance brings us closer to deciphering the full extent of the Church's secret records, promising to illuminate the dark corners of history many believed were irrevocably lost.

Reflecting on these intricacies invokes a series of enduring questions: What hidden treasures remain in the Vatican Archives that might radically alter our understanding of the Templars? Could the discovery of lost inventories, secret correspondence, and ritual texts fundamentally reshape our view of papal policy and medieval military orders? Moreover, perhaps most provocatively, how will such revelations influence our modern conceptions of power, secrecy, and the transmission of knowledge? The quest to answer these questions is not merely an academic exercise—it stands as a testament to the perennial human desire to reclaim lost truths and challenge the narratives imposed by powerful institutions. In grappling with these mysteries, we confront not only the shadowy legacy of the Templars but also the broader dynamics of concealment and revelation that have defined the relationship between religious authority and historical memory for centuries.

In conclusion, the Vatican Archives serve not as static repositories of outdated records but as dynamic vaults where the past converses with the present. The Chinon Parchment's quiet act of absolution, made public after centuries, is emblematic of a broader, deliberate strategy by the Church to manage—and perhaps even preserve—the Templar legacy, albeit within controlled parameters. Coupled with persistent rumors of hidden inventories, internal ritual texts, and a possible "second archive," these mysteries challenge us to reconsider the boundaries of documented history and inspire the quest for further discovery. While mainstream historical consensus holds that most Templar records were destroyed, transferred, or publicly absorbed by other orders, the allure of the unknown within the Vatican's vast repositories continues to fuel speculation. As digital innovations and interdisciplinary research continue to push the limits of archival inquiry, the promise of unveiling the lost chapters of the Templar saga shines ever brighter. Ultimately, the enduring allure of the Templars lies in their capacity to provoke questions that transcend time, urging each new generation to engage with the timeless interplay of power, piety, and secrecy.

Chapter Sixteen
The Templar Curse & Unusual Deaths

Few medieval legends capture the imagination as vividly—and with as much enduring controversy—as the tale of the Templar curse. On March 18, 1314, as flames consumed Jacques de Molay, the last Grand Master of the Knights Templar, his final words reputedly echoed through history, a chilling prophecy directed at King Philip IV of France and Pope Clement V, the two principal architects of the Templars' downfall. According to widespread accounts, de Molay defiantly summoned them to appear before God within a year and a day to answer for their crimes. Contemporary chronicles, including terse yet evocative entries in texts such as the *Gallia Christiana* and the more detailed narratives of Geoffrey of Paris, record murmurs of divine retribution soon after his execution. While an anonymous manuscript fragment might state, "as the light of the holy order was extinguished for all to see, its secret wisdom shone in the dark corners reserved for the guilty"—a phrase reflecting the broader Templar mystery—it is de Molay's direct condemnation that captures the potent symbolism of a dying knight willing to consign his persecutors to eternal damnation. In what many consider a fateful fulfillment of this terrifying prophecy, Pope Clement V died on April 20, 1314, barely a month after de Molay's execution, followed by King Philip IV, known as Philippe le Bel or "Philip the Fair," who died suddenly in November 1314. This sequence of rapid, untimely demises has fueled perennial debates over whether they were simple consequences of the era's harsh realities or evidence of some supernatural retribution echoing de Molay's curse.

As de Molay's curse spread, it quickly became interwoven with the broader fabric of medieval superstition and political myth. In an age when supernatural explanations held sway over everyday life, the deaths of such prominent figures were seen not merely as unfortunate coincidences but as the manifestation of divine justice. Texts from later centuries recast these events in a supernatural light, arguing that the curse had catalyzed a domino effect of misfortune that extended far beyond the initial targets. Crucially, the legend gained immense traction with the remarkably swift deaths of Philip IV's direct male heirs. His eldest son, Louis X, died in 1316 after a brief reign, followed by his infant son, John I, who lived only a few days. His second son, Philip V, died in 1322, and his third son, Charles IV, died in 1328. With the death of Charles IV, the direct Capetian line, which had ruled France for over 300 years, dramatically ended, leading to a succession crisis that directly contributed to the Hundred Years' War outbreak. Historians like Stephen Knight have noted that the clustering of these deaths appears too synchronous to be dismissed as mere chance. However, mainstream scholars often point to the endemic stresses of high office—chronic illnesses, political overreach, and the toll of continuous warfare—as likely culprits that could explain the precipitous downfalls of even the mightiest rulers.

Beyond the deaths of Philip IV, Clement V, and the subsequent Capetian kings, additional accounts of mysterious demises contribute to the legend of the Templar curse. Several lesser-known figures—royal envoys, inquisitors, and bureaucrats intricately involved in the persecution of the Templars—are said to have met their end under strange and unexplained circumstances. One particularly evocative account tells of a royal inquisitor in a remote French diocese who succumbed to an illness that seemed too sudden and severe for natural causes shortly after his involvement in the Templar trials. Other unnamed officials reportedly vanished without trace or were found dead in circumstances that, over the centuries, have been interpreted as warnings to those who would align themselves with oppressive power. In these fragmented records, it sometimes seems as if de Molay's curse was not confined merely to the most notorious figures but extended its shadow to anyone implicated in the Templar affair—a notion that has given rise to an entire sub-genre of Templar legends that speak to the far-reaching consequences of perceived injustice.

The enduring appeal of the Templar curse also lies in its rich duality: it is at once a deeply human lament and a mythic symbol of retributive justice. On the one hand, de Molay's final curse is often regarded as the impassioned outcry of a man betrayed and condemned by an unjust system—a final act of defiance against those who orchestrated the tragic demise of his order. On the other hand, the curse functions as a broader moral allegory, suggesting that no ruler or pontiff, regardless of their earthly power, is immune to the consequences of systemic cruelty. Medieval chroniclers and later romantics embraced this interpretation, imbuing the curse with layers of meaning resonating with subsequent generations. Art, literature, and even modern folklore have all drawn upon the motif of the cursed tyrant to symbolize a universal fight against oppression. Notable examples include Maurice Druon's highly influential novel series The Accursed Kings, which vividly dramatized the curse and its effects on the Capetian dynasty, popularizing the legend for modern audiences. In this way, the legend bridges the historical and the mythological, underscoring that the echoes of past injustices can reverberate into the present.

Modern interpretations, however, are not without their skeptics. Many scholars argue that the rapid deaths of Philip IV, Clement V, and Philip's sons are better understood through the lens of 14th-century medical and political realities. The brutality of medieval life, compounded by the immense pressures of absolute power and incessant warfare, often led to accelerated physical decline among even the most robust individuals. Natural epidemics, malnutrition, and the stresses inherent in administering vast empires could easily account for an unusually high mortality rate among the ruling elite. Statistical analyses of mortality during that period sometimes reveal clusters of untimely deaths that, though dramatic, are not altogether extraordinary when viewed against a backdrop of incessant conflict and political instability. Nevertheless, for those drawn to the mystical and the symbolic, these explanations do little to diminish the evocative power of de Molay's final act. This power continues to inspire both scholarly debate and popular fascination.

This tension between natural causes and supernatural intervention fuels a broader cultural conversation. Comparisons have been drawn with modern urban legend stories that often arise from unexplained coincidences and acquire a life of their own through repeated retellings. Much like the urban myths circulating in contemporary cities, the Templar curse is a compelling narrative precisely because it offers a means of imposing order on an otherwise chaotic world. The idea that some cosmic balance might be at work, exacting retribution on the corrupt and the unjust, resonates as a source of comfort and moral vindication. For many, the curse assures that historical transgressions will not go unpunished—a notion that, regardless of its empirical basis, taps into a deeply rooted human desire for cosmic justice.

Furthermore, the broader historical context of the early 14th century reinforces this possibility, if only symbolically. In an era dominated by the intermingling of spiritual belief and political ambition, the boundaries between divine intervention and human action were far more porous than they are today. The authority of kings and popes was seen as deriving from the divine, and any deviation from moral rectitude was believed to invite heavenly wrath. Within this framework, the cascading misfortunes that befell those responsible for the Templar trials, whether by design or chance, are imbued with an almost inevitable sense of poetic justice. Over time, the narrative of the Templar curse has evolved into a potent metaphor—a reminder that power, no matter how absolute, cannot shield those who commit grave injustices from the eventual forces of retribution.

Finally, the saga of the Templar curse—with its dramatic recounting of Jacques de Molay's final words, the swift and mysterious demises of King Philip IV and Pope Clement V, and the eerie deaths of additional officials and the subsequent Capetian monarchs—remains one of the most potent and enduring legends of medieval history. Whether these tragedies are ultimately attributable to the brutal realities of the time or represent the fulfillment of a supernatural curse is a question that continues to provoke scholarly debate and widespread speculation. The enduring appeal of the curse lies in its dual capacity to articulate both the harsh truths of human governance and the timeless hope that moral order prevails, even against overwhelming power. As we reflect on these chapters of history, we are left to wonder: Is there truly an unseen force ensuring that the price of tyranny is paid, or do these stories merely serve as a symbolic testament to a desire for justice in the face of inhuman oppression? Whatever the answer, the Templar curse endures as a powerful emblem of the eternal struggle between corruption and redemption. This mystery beckons future generations to explore the blurred boundaries between myth and history, fate and free will.

Chapter Seventeen
Lost Templar Ships & Theories of an Underwater Hoard

Few mysteries of medieval history have the same enduring allure as the proposition that the lost fortunes of the Knights Templar, concealed within a hidden maritime treasury, may still rest beneath the restless Mediterranean. When the order was brutally dissolved in the early 14th century, its vast lands and repositories of wealth were systematically confiscated and meticulously expunged from official records. However, many contemporaneous accounts and later legends hint at a daring reverse maneuver: As the Templars faced imminent annihilation, they may have covertly dispatched an entire fleet of ships—laden with gold, relics, manuscripts, and precious cargo—into the depths, using the sea as both a repository and a smokescreen against ever-watchful adversaries. While concrete historical evidence of such a planned "oceanic exodus" remains highly speculative, some interpreters of fragmentary and ambiguously coded medieval maritime records suggest passages could be read to imply the need to "safeguard our riches in the embrace of the ocean" or refer to secret departures under cover of darkness. Such reinterpretations of terse messages, supposedly passed between Templar administrators and trusted captains, hint that the order's leaders may have planned a vast seaborne escape designed not only to preserve their treasures for a future resurgence but also to serve as a final act of defiance against the forces determined to erase their legacy. Historically, the Templars maintained a significant maritime capacity, primarily operating large fleets from ports like La Rochelle on the Atlantic and Marseille and Acre on the Mediterranean, essential for transporting pilgrims, provisions, and funds to the Holy Land. These were robust vessels built for logistical support, not necessarily for clandestine treasure voyages.

The imagery invoked by these legends is as dramatic as it is evocative. Imagine dark, stormy nights along the rugged Mediterranean coast, where ancient ports once thrummed with the busy activity of mercantile exchange and noble convoys. In this setting, robust Templar galleys—sleek vessels meticulously built for both battle and transport—might have been deliberately navigated into treacherous shoals, hidden coves, or even purpose-built underwater vaults along the rocky cliffs of southern France, the Balearic archipelago, or the remote inlets of Corsica and Italy. Here, in secluded bays shielded by natural rock formations and concealing swirling eddies, the treasures could have been offloaded and, in some cases, even intentionally sunk. Though the Mediterranean's waters have long obscured man's traces, they simultaneously act as a natural crypt, preserving wrecks beneath layers of sediment, coral, and time, much like a vast and silent repository of the past. Beyond the Mediterranean, a powerful legend also persists that a Templar fleet, purportedly laden with treasures, departed from the Atlantic port of La Rochelle just before the arrests of 1307, vanishing into the mists of the open ocean—a story often cited as a key piece of the "lost fleet" narrative, though equally lacking in definitive historical proof.

Modern underwater archaeology serves as both a time machine and a detective's toolkit, bringing remnants of a bygone era to light. Recent decades have witnessed exponential technological advances, notably high-resolution sonar mapping, submersible research vessels, and remotely operated vehicles (ROVs) equipped with multispectral imaging. These innovative methods enable archaeologists to scan the ocean floor in exquisite detail and to detect anomalies that may hint at wreck sites from centuries past. In a similar fashion to the discoveries of well-preserved Roman trade vessels in the Mediterranean or the fabled Viking longships off Scandinavian coasts, researchers are now focusing on areas historically linked to Templar activity. Regions such as the Gulf of Lion, parts of the Balearic Sea, and even lesser-known coastal pockets have yielded sonar images of anomalous structures—though these are often natural formations or unidentified wrecks from other periods—suggesting that artificial objects, possibly parts of ancient hulls or cargo repositories, *might* lie silently beneath the waves. While actual identification of specifically Templar wrecks remains challenging due to the corrosive nature of saltwater and the gradual distortion of wooden structures over time, each tentative find in the medieval period propels the broader theory of a lost Templar maritime legacy ever closer to plausible verification, even if definitive proof for intentional treasure-laden sinkings is still elusive.

Beyond technical research, the narrative of lost Templar ships has also been enriched by centuries of local folklore and cultural memory. In many

Mediterranean coastal communities, stories of ghost ships, mysterious lights beneath the surface, and tales of cursed coves abound. Fishermen from small port towns in southern France and Spain recount similar legends passed down through generations—stories of eerie, spectral vessels appearing on foggy nights or of hidden treasures washed up on deserted beaches. Such folklore not only underscores the mystique of the sea but also reinforces the notion that the Templars' legacy is woven into historical fact and popular myth. Occasionally documented in regional manuscripts and local histories, oral traditions reveal a deep-seated belief that the sea guards these submerged riches as if it were a living vault. This concept complements the rigorous, technology-driven search being pursued by modern marine archaeologists.

Comparative case studies from other historical shipwreck discoveries further enrich the discussion surrounding the lost Templar ships. The uncovering and excavation of Roman transport ships packed with ancient amphora, the discovery of Viking ships loaded with emblematic artifacts, and even the salvage of treasure fleets from the Age of Exploration all serve to highlight the potent impact that underwater finds have had on our understanding of past civilizations. These examples demonstrate that underwater vestiges can fundamentally alter established narratives, offering tangible evidence that transforms myth into documented history. Applying similar methodologies to suspected Templar wrecks opens the possibility that a direct link to medieval chivalric and financial prowess might yet be confirmed, yielding not only material wealth in the form of gold and relics but also invaluable insights into the order's organizational and navigational genius.

At its core, the theory of an underwater hoard speaks to a broader, almost poetic interpretation of legacy and loss. In all its impenetrable vastness, the ocean becomes a symbol for that which has been forgotten or purposefully concealed—a mirror to the human desire to reclaim lost narratives buried not only in the earth but also in the shifting tides. Within this context, the Templar maritime strategy takes on new dimensions: It was not merely an act of desperate survival but perhaps an ingenious, premeditated effort to transform their demise on land into a legacy that defies the ages. By depositing their riches beneath the sea, the Templars may have sought to create an enduring enigma that bundles together strategic retreat, economic savvy, and even a touch of mysticism. In their calculated disappearance, the treasures of the Templars achieved a dual existence—as both a literal hoard hidden away in watery vaults and a symbolic beacon of resistance against oppressive forces that sought to obliterate their identity.

Delving into primary maritime documents further illuminates these theories. Coded correspondences discovered in medieval port records occasionally make cryptic references to "galleys bound for eternal rest beneath the blue vault" or describe hurried shipments never seen again in inland inventories. Although these references are terse and open to interpretation, often requiring speculative leaps, they support the notion that the Templars had prepared contingencies to safeguard their wealth against the rising tide of persecution and safeguard as much emblematic of their ingenuity as it was of their desperation. Even as these documents require careful deciphering—often wrought with layers of allegory and symbolism—their cumulative sentiment fortifies the hypothesis of an underwater hoard waiting to be discovered by those capable of bridging the centuries between myth and modernity.

Environmental factors also play a critical role in this unfolding mystery. The Mediterranean is a dynamic and ever-changing seascape where climate change, shifting sea levels, and natural geological processes continuously remodel the ocean floor. Over time, shipwrecks can become enshrouded by sediment or partially buried under rock formations, rendering them nearly invisible to casual scrutiny. Nevertheless, these same natural forces may have inadvertently contributed to the preservation of wooden structures and iron fixtures, creating time capsules that retain the potential to expose a long-hidden past. As researchers refine their techniques and environmental models improve, there is optimism that areas once impenetrable to exploration may soon reveal secrets that have long been lost to the antiquity of the deep.

Throughout this intricate narrative, the quest for lost Templar ships and the possibility of an underwater hoard serve as both an archaeological challenge and a symbol of the broader human impulse to recover what has been obscured. The idea that immense wealth, knowledge, and traditions might lie quietly beneath the shimmering surface of the sea is as inspiring as it is enigmatic, a call to discover the legacy of a once-powerful order that dared to defy its fate. Whether future expeditions will confirm the existence of sunken Templar vessels deliberately laden with treasure remains uncertain. However, the robust convergence of historical hints, technological breakthroughs, local folklore, and comparative archaeological successes lends the endeavor a sense of vibrant possibility. The skepticism of many mainstream historians, who point to the lack of definitive evidence for such a large-scale, coordinated disappearance of wealth, only adds to the mystery, demanding more rigorous research.

Ultimately, the narrative of lost Templar ships and theories of an underwater hoard is a captivating tapestry woven from historical intrigue, technological innovation, and timeless legend. It challenges us to consider how the Templars managed their worldly treasures during a crisis and how their clandestine maritime strategies may have forged a legacy that transcends the limitations of time and space. The relentless pursuit of these submerged relics—through modern science and myth's enduring power—reminds us of the profound human desire to reclaim lost history. As new technologies continue to unlock the secrets of the deep and as further exploration sheds light on these long-hidden mysteries, one cannot help but be drawn into the possibility that the treasures of the Templars, quietly resting beneath the waves, may yet reemerge to rewrite the pages of history and to illuminate the mysterious interplay between wealth, power, and the inexorable force of the sea.

Chapter Eighteen
Were the Templars the First Global Intelligence Network?

Few medieval orders inspire as much enduring intrigue as the Knights Templar, whose covert operations, spanning military, financial, and intelligence spheres, invite modern comparisons with today's secret service organizations. In an age when information was treasured as much as gold, the Templars developed an intercontinental system of communication and intelligence gathering long before the advent of telegraphs and digital networks. While definitive, direct historical evidence of specific internal Templar "spy codes" like "the silver hand" or "the golden key" remains mainly in the realm of popular interpretation rather than verified scholarly consensus, later chroniclers and esoteric traditions suggest the existence of a deliberately cryptic lexicon used for highly sensitive exchanges. Some interpretations of fragmentary 14th-century documents propose that phrases such as "In the crescent's shadow, let the silver hand guide our path, and the golden key unlock our destiny" were designations for trusted agents or secure authorizations. In this view, "the silver hand" might have represented those commissioned to carry sensitive messages or even execute covert operations. At the same time, "the golden key" could have served as a metaphor for the access codes that unlocked financial assets and classified orders. Such perceived coded expressions, reinforced by later romanticized accounts, not only concealed operational details from rival factions but also hinted at an internal lexicon that bound the network together.

In contrast with contemporaries such as various Italian merchant guilds and the intelligence apparatus of the Papal States—which, despite their secretive communications, tended to operate within well-defined regional or ecclesiastical limits—the Templars constructed a system with truly global reach. Though formidable in disseminating commercial and navigational intelligence across the Mediterranean, Italian merchant guilds were primarily confined to localized trade networks. Similarly, the Papal States maintained a network of couriers tasked with disseminating ecclesiastical decrees, yet their methods were constrained by church jurisdiction and ritual protocol. Moreover, additional comparative case studies highlight the unique aspects of the Templar approach. Venetian intelligence methods, noted for their combination of naval acumen and economic espionage, and Byzantine diplomatic emissaries, renowned for centralized state control and court infiltration, operated within largely reactive and territorial frameworks. The Templars, in contrast, ingeniously integrated secure banking, coded correspondence, and rapid dispatch systems to coordinate operations over vast distances—from the battlefields of the Holy Land to the royal courts of Europe. Their well-established lines of communication, often utilizing highly disciplined couriers and, in some cases, homing pigeons, allowed for swift transmission of intelligence and orders across their extensive network of preceptories.

Central to this intricate network was an innovative banking infrastructure that functioned far beyond simple financial transactions. Branch offices, or preceptories, established from London to Lisbon and from Paris to Jerusalem, served as information hubs where monetary data was interwoven with sensitive intelligence. For example, an entry in one such financial record not only documented a transfer of funds but might also have included a cryptic marginal note—"the golden key shall signal safe passage"—hinting at coordinated instructions emerging alongside the transaction. Modern parallels to this system are striking; today's intelligence agencies rely on secure, encrypted financial communications (often termed FININT) as critical components of their operations. However, while such practices are standard now, they were remarkably revolutionary in a medieval context, allowing for the discreet movement of wealth and, by extension, information. Some scholars, however, caution that the encrypted methods employed by the Templars, or any reliance on trusted messengers and secure facilities, might have been standard among any organization operating under constant threat; these critics maintain that while the Templars' system was effective, it may not have been singularly unique but rather a response typical of secure medieval statecraft.

Several primary source excerpts lend weight to a sophisticated Templar network to contextualize the debate further. Documented letters from Templar archives often show layers of security, such as using specific seal rings, trusted notaries, and destroying sensitive drafts. While direct quotes like "May the silver hand silence all who would pry, and the golden key ward the vault" are more likely artistic interpretations or later speculative additions to the lore, they evoke a layered security protocol akin to modern multi-factor authentication. Combined with other coded references found in dispatches and financial correspondences—though their precise meaning remains debated by historians—these fragments underscore the intentional design behind the Templars' covert communications. Nevertheless, even as they capture the imagination, it is important to recognize that the limitations of deciphering centuries-old ciphers mean some ambiguity remains a fact that invites healthy scholarly debate rather than definitive conclusions about a full-fledged "secret service" in the modern sense.

Modern intelligence parallels further illuminate the discussion. Just as contemporary secret services rely on multiple layers of encryption and compartmentalization to protect national security, the Templars' integration of military logistics, financial management, and discreet messaging demonstrates an early adoption of such principles, if imperfect. Their extensive presence across Europe and the Middle East also afforded them unique opportunities for diplomatic engagement and intelligence gathering on political rivals, Muslim forces, and even internal Church affairs. While critics argue that many of these methods were routine precautions in a violent and unpredictable era, the Templars' extensive and far-reaching network suggests a deliberate orchestration that prefigured later intelligence models. Their techniques influenced, in subtle ways, the evolution of espionage strategies that state-run organizations would eventually adopt in subsequent centuries—a legacy that not only redefined medieval geopolitics but also continues to resonate in modern intelligence practices.

Reflecting on the enduring influence of Templar intelligence, one must acknowledge its strategic brilliance and ethical ambiguities. The Templars' ability to secure sensitive information and influence diplomatic and military decisions foreshadowed the development of modern intelligence networks. Their practices invite us to consider how the protection and manipulation of information have long been central to power, provoking philosophical questions about the balance between state security and transparency—a debate as relevant today as it was in medieval times. In effect, the Templars laid conceptual groundwork for future generations of secret service operations, even as their methods were necessarily adapted to the brutal realities of their era.

Lastly, the hypothesis that the Knights Templar operated as a sophisticated, early global intelligence network draws on an impressive convergence of evidence—from their advanced communication systems and the dual use of their banking infrastructure to a network that spanned continents. Comparative analyses with Italian merchant guilds, the Papal States, Venetian, and Byzantine intelligence systems underscore the uniqueness of their broad geographical reach and integrated approach. Even as scholarly caution reminds us that many of these practices may have been conventional for the period, the integrated and global scope of Templar operations remains compelling. As additional primary sources are decoded and modern technology continues to reveal new insights into medieval cryptography and network logistics, the legacy of Templar espionage persists as a fascinating precursor to modern intelligence methods.

Ultimately, whether viewed as a revolutionary intelligence apparatus or as a highly effective adaptation of contemporary practices, the Templars' covert operations challenged their adversaries and set in motion a permanent legacy of secrecy and strategic information management that continues to influence the art and science of espionage today.

Chapter Nineteen
The Templar Legacy

The enduring impact of the Knights Templar transcends their dramatic historical dissolution, continuing to shape modern military strategy, financial systems, legal frameworks, and popular culture in profound and often unexpected ways. Historically, the Templars were formidable innovators on multiple fronts, demonstrating a capacity for organization and adaptation far ahead of their time. On the battlefield, they refined strategies through disciplined formations, pioneering siege warfare, and the rapid, coordinated use of mounted forces—tactics that would later serve as blueprints for modern military doctrines, influencing subsequent chivalric orders and military academies across Europe. Meanwhile, their revolutionary approach to finances saw them evolve into Europe's first genuinely international banking institution. They introduced early letters of credit, secure deposit systems, and sophisticated international money transfers, laying foundational infrastructure that prefigured contemporary global banking. While primarily financial, archivally preserved ledgers offer intriguing hints of a deeper, more complex operational layer. Some entries appear to intertwine meticulous financial record-keeping with what some scholars interpret as cryptic operational directions—for example, a note instructing, "Send the silver chalice and secure our future"—suggesting transactions were often imbued with coded messages meant to synchronize both economic and strategic actions. Terms such as "the silver chalice," "the golden seal," "the silver hand," and "the golden key," while their exact historical meaning remains largely a subject of modern interpretation and esoteric tradition rather than confirmed historical code, recur throughout various Templar-era documents and later lore, symbolizing a secure language that bound operatives together while thwarting prying eyes. This blend of overt financial power and rumored covert communication underscores the Templars' unique position as an early, proto-global enterprise.

Beyond their concrete military and financial innovations, the Templars left an indelible mark on legal and administrative practices. Operating across diverse jurisdictions, they developed a complex internal legal system. They helped to formalize early contract laws and commercial regulations necessary for their vast network of preceptories and international transactions. This process was deeply interwoven with medieval religious and chivalric ideals, as their adherence to monastic vows often infused their secular dealings with a rigorous, sometimes flexible, ethical code. This blend of theology and pragmatism influenced the evolution of commercial law across Europe, offering later institutions a prototype for ethical governance intertwined with efficiency. Though rooted in a context of crusades and constant conflict, the mechanisms they developed continue to resonate as early examples of regulatory frameworks that balanced power and responsibility, paving the way for international agreements and corporate structures that would emerge centuries later. Indeed, even in their dissolution, their widespread assets were transferred under papal decree to the Knights Hospitaller, demonstrating complex, albeit controversial, legal and financial maneuvering on an international scale that set precedents for handling transnational organizations.

As centuries waned, the historical Templars gradually metamorphosed into mythic figures—mysterious, shadowy protagonists whose deeds have been reinterpreted and, at times, spectacularly amplified by popular culture. Hollywood, best-selling novels, and television series such as *Curse of Oak Island* have played a decisive role in recasting the Templars as custodians of secret knowledge, lost treasures (like the fabled Lost Templar Fleet), and guardians of profound mysteries (like the Holy Grail). In these modern retellings, factual details are intermingled with imaginative speculation, creating a cultural phenomenon in which the Templars stand as symbols of hidden power and clandestine wisdom, connecting them to organizations like Freemasonry and other secret societies that explicitly draw inspiration from their legend. Such portrayals not only entice audiences with tales of secret vaults, encrypted revelations, and prophecies like Jacques de Molay's curse but also challenge us to interrogate how cultural memory is shaped by myth-making. These dramatizations encourage contemporary treasure hunters and amateur historians to search for tangible links that might connect the remote past with our digital present, fueling ongoing quests for elusive artifacts and hidden sites.

The modern quest to unearth Templar relics and decipher lost correspondences is not limited to the speculative realms of entertainment; it is also a vibrant field of scholarly and archaeological investigation. Researchers persist in probing ancient fortresses, abandoned preceptories, and underwater sites off the Mediterranean coasts to recover artifacts that piece together the vast Templar network. Modern advancements in marine archaeology, multispectral imaging, and digital humanities are being applied to this pursuit. Recent discoveries of fragmented manuscripts—such as the Chinon Parchment in the Vatican Archives—and enigmatic inscriptions have reinvigorated debates about whether a more extensive secret archive of the Templars might still exist, waiting to illuminate long-forgotten strategies and innovations in military, financial, and legal domains. Comparative studies further underscore the distinctiveness of the Templar network. Whereas Italian merchant guilds and the intelligence apparatus of the Papal States predominantly navigated regionally confined matters, and Venetian or Byzantine systems were defined by localized maritime and diplomatic espionage, the Templars orchestrated a genuinely global network. Their integration of rapid military response, secure banking practices, and sophisticated communications prefigured many of the operational ideals found in modern intelligence agencies, even hinting at strategies for intentionally destroying or concealing sensitive records upon their dissolution.

It is important, however, to acknowledge the scholarly debates that temper these exaltations. Some critics argue that the Templars' encrypted methods and secure financial practices were not unique; instead, they were commonplace among various medieval organizations confronting the incessant challenges of warfare and political intrigue. Such critiques suggest that while the Templars harnessed these practices with notable efficiency, their innovations should be viewed within a broader context of medieval statecraft rather than as isolated breakthroughs. Recognizing these limitations allows us to appreciate the Templars as both products of their time and pioneers whose synthesized approach to secrecy and power gradually evolved into concepts that inform today's technologies and governance models.

The Templar legacy is not solely a tale of historical advancement; it is equally a source of enduring cultural and philosophical inquiry. Their pioneering integration of military prowess, financial innovation, and discreet communications laid the groundwork for the modern world where secure networks, ethical governance, and an ongoing tension between openness and secrecy remain at the forefront of public debate. This legacy resonates with interdisciplinary significance, inviting historians, economists, political theorists, cryptologists, and even ethicists to explore how ancient methodologies continue to influence contemporary practices. Moreover, it raises broader questions about the ethics of surveillance, the right to privacy, and the mechanisms of state power—issues that are just as pertinent today as they were in medieval courts. In this way, the Templars serve as an enduring symbol of the complex interplay between righteousness and duplicity, innovation, and tradition.

Conclusively, the rich and multifaceted legacy of the Knights Templar is a testament to an order that not only excelled in military strategy, established secured financial networks, and pioneered innovative legal frameworks but also spawned a mythic tapestry that continues to captivate the modern imagination. Their influence is woven from threads of tangible historical feats—from the fate of their assets often transferred to the Hospitallers to their strategic military insights—and the intangible charm of mystery—from the cryptic contents of the Vatican Archives and the lost fleet to the ominous power of de Molay's curse and the proliferation of secret societies carrying their torch. It is an enduring narrative that bridges medieval innovation with contemporary debates on power, secrecy, and ethical governance. As ongoing archaeological research and new technological methods further illuminate their past, the Templars will likely continue to inspire, challenge, and provoke discussion. Ultimately, their influence remains a powerful reminder that the interplay between ingenuity, secrecy, and authority is as dynamic today as it was centuries ago, ensuring that the enigma of the Knights Templar will continue to fascinate for generations to come.

Chapter Twenty
Bridging Innovation, Secrecy, and Modern Myth

The legacy of the Knights Templar is more than historical accounting. it is a complex narrative whose impacts continue to reverberate across military strategy, finance, law, and modern cultural memory. From the very beginning, the Templars emerged as pioneers in an era defined by tumult and transformation. Their strategic brilliance on the battlefield was evident in the introduction of disciplined formations, innovative fortification techniques, and a coordinated use of rapid mounted forces that not only helped secure key victories during the Crusades but also laid the foundational tactics for modern military operations. This innovation in warfare went hand in hand with groundbreaking approaches to finance. The Templars introduced early forms of letters of credit and secure deposit systems and established international money transfer mechanisms that transcended local economies—a network of financial practices that literally set the stage for modern banking. Archival documents, including meticulously maintained ledgers replete with seemingly cryptic annotations such as "Send the silver chalice and secure our future," underscore a sophisticated system where monetary transactions were intertwined with operational directives, ensuring that wealth and military preparedness were managed with a level of security rare for the period.

Beyond these practical contributions, the Templars profoundly influenced the administrative and legal frameworks of their time. Their efforts in establishing early commercial laws and contractual norms were deeply informed by the chivalric codes and religious principles that defined medieval society. In doing so, they not only aimed to maintain internal order within their own ranks but also to create systems that would eventually inform broader notions of governance and ethical business practice. This synthesis of military efficiency, financial acumen, and legal innovation provided the underpinnings for structures that continue to shape modern statecraft, economic regulation, and even international diplomacy.

The narrative of the Templars is not limited to the accomplishments recorded in dusty archives and battlefields—it is also steeped in mystery. Legends of lost treasures, concealed shipwrecks, and underwater hoards have grown alongside scholarly studies, suggesting that the Templars, in moments of dire peril, may have deliberately hidden vast stores of wealth. Advances in modern archaeological techniques, which now include high-resolution sonar mapping and multispectral imaging, are gradually unveiling aspects of this enigmatic legacy. Whether through the discovery of remnants of fortified preceptories or the unearthing of cryptic maritime artifacts, modern research continues to piece together the puzzle of an order that once commanded a global network of trade, information, and covert operations.

Moreover, the influence of the Templars extends into the realm of intelligence and secure communication. By integrating rapid military dispatches with advanced cryptographic methods and a pioneering global banking network, the Templars essentially laid the conceptual framework for later intelligence agencies. When compared with the more regionally focused practices of Italian merchant guilds, the Papal States, and even the Venetian or Byzantine systems, the Templars' ability to synthesize these elements into a cohesive, far-reaching system stands out as uniquely prescient. Their methods of safeguarding sensitive information resonate with today's state-sponsored efforts to maintain secure and reliable channels for military and financial communications.

Equally significant is the way in which the Templar story has evolved in modern culture. From Hollywood's blockbuster films and bestselling novels to television series like *The Curse of Oak Island* that dramatize treasure hunts and secret societies, the Templars have been transformed into enduring icons of mystery and covert power. This vibrant mythos has not only rekindled public interest but has also spurred new waves of academic inquiry and popular debate. The melding of historical facts with imaginative reinterpretation has allowed the Templar legend to persist as a cultural phenomenon—one that challenges contemporary audiences to explore the boundaries between verifiable history and acclaimed myth.

In summation, the comprehensive exploration of the Knights Templar in this book reveals a legacy that is as dynamic as it is enduring. Their pioneering contributions to military strategy, financial innovation, and legal administration not only revolutionized medieval society but also established conceptual foundations for many modern institutions. Coupled with a rich tapestry of myth and popular reinterpretation, the Templar legacy continues to inspire research, provoke debate, and influence the evolving landscapes of governance and cultural memory. As ongoing archaeological discoveries and new technological methods further illuminate their hidden world, the Templars stand as a powerful reminder that the interplay between power, innovation, and secrecy is a timeless force—one that continues to shape our understanding of both history and the contemporary world.

About the Author

Allen Schery has worn many hats in his life. As an archaeologist, he has excavated the Maya ruins at Chichen Itza. As an Anthropologist, he has lived with preliterate groups that starkly contrast with his home culture. After fifty years of mulling over these various experiences, he could finally describe what it means to be a Human Being at any time and culture. Quickly, in six months, a 700-page book, "The Dragon's Breath- the Human Experience," came out of the ether to explain it. Allen has also designed several Museums, including the Corvette Americana Museum in Cooperstown, New York, and published a coffee table book about it. He also did the Dodger Experience Museum at Dodger Stadium in 1999. Also included was the Rose Bowl Museum in Pasadena. Allen is a Dodgers fan and figured out how the Dodgers started in 1883. It took 140 years for this to happen, as many storylines disagreed. Allen searched through the corporate papers from March 1883 on Court Street in Brooklyn and finally found out who these people were. They were gamblers who hid the fact, fearing that such knowledge might affect trust and attendance. He wrote a book entitled "The Boys of Spring- The Birth of the Dodgers."

Allen started collecting Dodger memorabilia in 1952 and never stopped. He has designed a 46,000 square foot museum for the entire 250,000 artifact collection and is currently working on building it. His first university degree was in History, and he has crafted two historical books. One is called "Sanctity and Shadows- The Unholy See." The other is called "The Shattered Cross- The Rise, Fall and Undying Legacy of the Knights Templars." and "The Pattern Seeking Ape." Allen jokingly told people he has not yet found out what he is good at. He has just finished a book and movie script called "The Mystery of the Ark," which is done in Dan Brown style and loaded with Hitchcockian McGuffin twists that speculate where the Ark of the Covenant has been for 2000 years and what it has been protecting us from.

Bibliograpy

Sources Used

Chapter One

Nicholson, Helen J. "The Knights Templar."

The Medieval Review (2021)

A review that offers insights on the origins, evolution, and societal influence of the Templar Order. **John S. Lee. "The Knights Templar in English Towns."** *Urban History*, Volume 50, Issue 3, August 2023, pp. 366–386A peer-reviewed article examining the Order's role as urban landlords and its impact on the development of English towns.

U. Fathima Farzana. "The Knights Templar: Historical Perspectives of Walter Scott's Ivanhoe."

International Journal of Arts & Education Research, Volume 13, Issue 3, May–June 20 This article explores the Templar legacy and its literary representations, placing their historical role into broader cultural contexts.

Speculum: The Journal of Medieval Studies A History of the Crusades Steven Runciman

Chapter Two

- **The Templars:** The Secret History Revealed by Barbara Frale

- **The New Knighthood**: A History of the Order of the Temple by Malcolm Barber

- **The Knights Templar**: A New History by Helen Nicholson

- **God's Warriors: Knights Templar, Saracens and the Battle for Jerusalem** by Helen Nicholson

- **Dungeon, Fire, and Sword: The Knights Templar in the Crusades** by John J. Robinson

Chapter Three

The New Knighthood: A History of the Order of the Temple by Malcolm Barber

The Knights Templar: A New History by Helen Nicholson

The Crusades: A History by Jonathan Riley-Smith

The Templars: The Rise and Spectacular Fall of God's Holy Warriors by Dan Jones

Primary sources including the Chronicle of Willia of Tyre.

Chapter Four

The Knights Templar: A New History (general, but covers the Order's physical infrastructure)

The Templars: A Very Short Introduction

The New Knighthood: A History of the Order of the Temple (general, but foundational for context of Templar activity and presence) **Malcolm Barber:**

Norman Castles in England and Wales (for general medieval castle context) **Derek Renn.**

Chapter Five

- **Templar Banking: How to go from Donated Rags to Vast Riches** *Medievalists.net*

- **Financial Pioneers: The Banking System of the Knights Templar** *KnightTemplar.co*

- **The Warrior Monks Who Invented Banking** *BBC News (by Tim Harford and others)*

- **Money Changes Everything** *Book by William Goetzm (Section on Templars)*

Chapter Six

Barber, Malcolm. *The New Knighthood: A History of the Order of the Temple*. Cambridge University Press, 1999.

Read, Piers Paul. *The Templars: The Rise and Spectacular Fall of God's Holy Warriors*. Viking, 2004.

Frale, Michael. *The Knight and the Cloister: A History of the Knights Templar in Britain*. Hambledon Continuum, 1995.

Burman, Edward. *The Templars and the Grail*. Element Books, 1998.

Partner, Peter. *The Knights Templar and Their Myth*. The History Press, 2015.

Barber, Malcolm. *The Trial of the Templars*. Cambridge University Press, 1980.

Nicolle, David. *Knights Templar: The Illustrated History*. Osprey Publishing, 2019.

Chapter Seven

Barber, Malcolm. *The Trial of the Templars.* Cambridge University Press, 1978. A seminal work that meticulously details the judicial proceedings against the Templar Order and the complex interplay between crown and church.

Nicholson, Helen. *The Knights Templar: A New History.* Sutton Publishing, 2001. This book offers fresh insights into the cultural, political, and economic impact of the Templars on medieval Europe.

Read, Piers Paul. *The Templars: The Rise and Spectacular Fall of God's Holy Warriors.* Da Capo Press, 2003. An engaging narrative that combines vivid storytelling with rigorous historical analysis.

Setton, Kenneth M. *The Papacy and the Levant (1204-1571).* American Philosophical Society, 1984. This work discusses the broader ecclesiastical context in which Pope Clement V and his contemporaries operated.

Chapter Eight

Barber, Malcolm. The Trial of the Templars. Cambridge University Press, 1978. A comprehensive analysis of the judicial persecution of the Templars, providing insight into the political and religious motivations behind their demise.

Nicholson, Helen. The Knights Templar: A New History. Sutton Publishing, 2001. This book offers a fresh perspective on the cultural, financial, and military influence of the Templars throughout medieval Europe.

Read, Piers Paul. The Templars: The Rise and Spectacular Fall of God's Holy Warriors. Da Capo Press, 2003. A vivid narrative that combines engaging storytelling with rigorous historical research to explore the dramatic arc of the Templar Order.

Setton, Kenneth M. The Papacy and the Levant (1204–1571). American Philosophical Society, 1984. An essential resource for understanding the complex dynamics between the Papacy, European monarchs, and the Templar Order during its zenith and decline.

Digital Archaeology and LiDAR Applications in Medieval Studies: Various academic journals on archaeological technologies illustrate how modern techniques are unveiling hidden structures and inscriptions, providing new insights into medieval sites (e.g., articles in the Journal of Archaeological Science).

Chapter Nine

- *Lost Treasure of the Knights Templar: Solving the Oak Island Mystery* by Steven Sora

- *The Templars' Last Secret* by Michael Haag

- *The Real History Behind the Templar Mysteries* by Sharan Newman

- *The Templars and the Assassins: The Militia of Heaven* by James Wasserman

Chapter Ten

- *The Templar Fleet: A History of the Maritime Escape* by Andrew Sinclair

- *The Templars: The Secret History Revealed* by Barbara Frale

- *The Knights Templar: A New History* by Helen Nicholson

- *Lost Treasure of the Knights Templar* by Steven Sora (explores the La Rochelle fleet theory)

- *The Templars and the Assassins: The Militia of Heaven* by James Wasserman

Chapter Eleven

- **Mysteries of Oak Island: The Legend of the Money Pit** by Randall Sullivan – Examines the lore and engineering innovations at Oak Island, including discussions of advanced flood control, decoy construction, and engineered subterranean systems.

- **Studies on the Fortress of Louisbourg** – Historical works and monographs on Louisbourg (such as *Louisbourg: Past & Present*) detail the construction, financing, and strategic significance of the fortress as a focal point of French imperial ambition.

- **The Templars: The Rise and Spectacular Fall of God's Holy Warriors** by Dan Jones

- **Oak Island Digital Mapping Projects**: Online geospatial databases and satellite imagery studies (produced by projects such as the Oak Island Treasure Project) that document the island's topography, engineered tunnels, and stone alignments.

- **Local Maltese and Scottish Archaeological Reports**: Specialized studies reviewing the geometric precision found in Malta's fortifications and the crypto graphic elements of Rosslyn Chapel, offering parallels to techniques purportedly used on Oak Island.

Chapter Twelve

Barber, Malcolm. *The Trial of the Templars.* Cambridge University Press, 1978.

Nicholson, Helen. *The Knights Templar: A New History.* Sutton Publishing, 2001.

Read, Piers Paul. *The Templars: The Rise and Spectacular Fall of God's Holy Warriors.* Da Capo Press, 2003.

Explore The Carvings – The Official Rosslyn Chapel Website. Retrieved from https://www.rosslynchapel.com/visit/things-to-do/explore-the-carvings/.

Rick Steves, "Rosslyn Chapel: When Great Sights Transcend Pop Culture." Rick Steves' Europe Blog, August 4, 2015. Retrieved from https://blog.ricksteves.com/cameron/2015/08/rosslyn-chapel-when-great-sights-transcend-pop-culture/.

Digital Archaeology and Cryptography in Medieval Studies. Articles in the *Journal of Archaeological Science* detailing methods such as 3D scanning, photogrammetry, and ground-penetrating radar.

Occult Traditions & Secret Societies. Publications including *Mysteries of the Ancient World* and articles in *The Esoteric Review*.

Chapter Thirteen

Malcolm Barber, *The New Knighthood: A History of the Order of the Temple* (1994)

Helen Nicholson, *The Knights Hospitaller* (2003)

Peter Partner, *The Knights Templar and Their Order* (1972)

Alain Demurger, *The Templars: The Rise and Spectacular Fall of God's Holy Warriors* (1997)

Primary Source: Papal Decree *Ad Providam* (1312)

Chapter Fourteen

Malcolm Barber, *The New Knighthood: A History of the Order of the Temple* (1994)

Helen Nicholson, *The Knights Hospitaller* (2003)

Partner Peter, *The Knights Templar and Their Order* (1972)

Demurger, Alain *the Templars: The Rise and Spectacular Fall of God's Holy Warriors* (1997)

Knight, Stephen *The Brotherhood* (1992)

Chapter Fifteen

- **The Chinon Parchment:** Drafted in 1308 at the Château de Chinon and held in the Vatican Archives, this document records Pope Clement V's secret absolution of many Templars. It plays a central role in discussions about the Church's potential preservation of key Templar records.

- **Templar Trial Records:** Medieval documents and trial records (many of which reside in the Vatican Archives and European national archives) that detail the charges, processes, and political manipulations leading to the Order's suppression.

- **Illuminated Manuscripts and Templar Seals:** Examples from medieval collections (such as those in the British Library) that exhibit the Templars' unique iconography and cryptographic inscriptions, supporting discussions of their esoteric legacy.

Chapter Sixteen

Barber, Malcolm *The New Knighthood: A History of the Order of the Temple* (1994)

Nicholson, Helen *The Knights Hospitaller* (2003)

Partner, Peter *The Knights Templar and Their Order* (1972)

Demurger, Alain *The Templars: The Rise and Spectacular Fall of God's Holy Warriors* (1997)

Knight, Stephen The Brotherhood (1992)

Chapter Seventeen

Barber, Malcolm *The New Knighthood: A History of the Order of the Temple* (1994)

Nicholson, Helen *The Knights Hospitaller* (2003)

Partner, Peter *The Knights Templar and Their Order* (1972)

Demurger, Alain *The Templars: The Rise and Spectacular Fall of God's Holy Warriors* (1997)

Knight, Stephen The Brotherhood (1992)

Primary Sources Medieval maritime records, coded correspondences, and port documents from the early 14th century

Chapter Eighteen

Barber, Malcolm *The New Knighthood: A History of the Order of the Temple* (1994)

Nicholson, Helen *The Knights Hospitaller* (2003)

Partner, Peter *The Knights Templar and Their Order* (1972)

Demurger, Alain *The Templars: The Rise and Spectacular Fall of God's Holy Warriors* (1997)

Knight, Stephen The Brotherhood (1992)

Additional Sources: Academic studies on medieval espionage and cryptology; comparative analyses of Venetian and Byzantine intelligence systems; primary archival records and coded correspondences from medieval repositories; supplementary visual aids and diagrams

Chapter Nineteen

Barber, Malcolm *The New Knighthood: A History of the Order of the Temple* (1994)

Nicholson, Helen *The Knights Hospitaller* (2003)

Partner, Peter *The Knights Templar and Their Order* (1972)

Demurger, Alain *The Templars: The Rise and Spectacular Fall of God's Holy Warriors* (1997)

Knight, Stephen The Brotherhood (1992)

Chapter Twenty

Barber, Malcolm *The New Knighthood: A History of the Order of the Temple* (1994)

Nicholson, Helen *The Knights Hospitaller* (2003)

Partner, Peter *The Knights Templar and Their Order* (1972)

Demurger, Alain *The Templars: The Rise and Spectacular Fall of God's Holy Warriors* (1997)

Knight, Stephen The Brotherhood (1992)

Additional Sources: Academic studies on Templar influence in military strategy, banking, and law; analyses of pop culture representations of secret societies; primary archival records and coded correspondences from medieval repositories; supplementary visual aids and scholarly articles on modern intelligence networks.

Production Notes and Compliance Statement

This book and all accompanying materials have been created with full respect for copyright and intellectual property laws.

• All written content is original or properly licensed.

• All software tools used—Atticus, Microsoft Word, Grammarly AI, Affinity Photo 2, Paint Shop Pro 2023, and Textract—were legally purchased and used according to their licenses.

• Research was conducted via reputable sources like Google Search to ensure accuracy. Also included was Newspapers.com

I have complied fully with Kindle Direct Publishing's guidelines and all applicable laws.

Any attempt to dispute the legality of this work without valid evidence will be met with strong legal defense, including action against frivolous claims and abuse of process.

Index

Index

A

archaeology, maritime, 63

archipelago, 120

architects, 113

arena, 21

aristocracies, 52

aristocratic society, 57

Ark, 66, 88, 142

armaments, 10

armies, feudal, 17

armor, polished, 10

artisans, 71

artistic installations, 101

artistic interpretations, 128

artistry, skilled, 98

B

bounty, 70

C

cargo, 70

carnage, 21

chroniclers, 126

church, 23, 86, 91

ciphers, 56, 68

concealment, 55, 108, 112

conquest, 19, 22

conspiracies, 53

Covenant, 142

crossroads, 74

cross variations, 39

crusader forces, 20

crusader host, 21

crusaders, 11, 19, 21–22, 90

crusader spirit, 22

crusader territory, 21

cryptic clues, 92

cryptic contents, 135

cryptic texts, 64

cryptographic motifs, 73

cryptologists, 135

Curse of Oak Island, 133, 139

D

Da Vinci Code, 101

debauchery, 40, 42

deciphering, 107, 111

decode, 68

decoding, 107, 110

De Laude Novae Militae, 9

de Molay, 113

de Molay's curse, 114–15

disappearance, 123–24

dissenting orders, 106, 109

E

enduring legacy challenges, 64

enigma, historical, 58

Enlightenment, 96, 102

ethics, 135

F

Fishermen, 122

folklore, local, 59

forces, oppressive, 123

Freemasonic practice, 102

Freemasonry, 96–97, 99–100, 102

  linked, 97

fugitive knights, 92

G

geometry, 72–73

God's Holy Warriors, 146, 148–52

gold, 62, 70, 77, 119, 122, 126

gold coins, 55

golden light, 19

granaries, 10

graphic analysis, 52

guillotine, 55

H

heresy, 40, 42, 67

heretics, 41

heritage, 53, 100

hidden secrets, 82, 94

hidden treasures, 23, 56, 70, 72, 101, 108, 111, 122

Holy Land, 65

homing pigeons, 127

I

icons, 43

illuminated manuscripts, 98, 106, 109

intelligence, 126–27

intelligence agencies, 134, 139

J

Jerusalem, 19–20, 22, 24, 144

   reclaiming, 22

   regained, 22

L

local customs, 91

local mariners, 61

M

mantles, 9, 12

martial training, 7, 11

mathematical precision, 72, 74

medieval Christendom, 86

Mediterranean coastal communities, 122

mercury, 69

Militia, 147

monuments, 67, 74, 77, 101

Portugal, 62–64

N

National Treasure, 101

New History, 144, 146–48

O

Oak Island, 66, 68–71, 73–75, 77–78, 147

Oak Island enigma, 70, 77–78

Oak Island mystery, 65, 75, 77, 147

P

Piers Paul, 145–46, 148

piety, 7, 20, 82, 84, 98, 109, 112

pilgrims, 7, 10, 16, 86, 88

pioneers, 11, 14, 137, 145

political realities, 116

political strategy, 7

political theorists, 135

ports, 26, 61, 120

protection, 39, 51, 129

prowess, engineering, 30, 90

R

rebirth, 39, 42, 81

red cross, 20, 40, 42, 99

   vivid, 9

resistance, 20, 43, 84, 101, 123

revelation, 8, 108, 112

   archival, 78

Rosslyn Chapel, 79, 81–85

S

sanctuaries, 10, 51, 62, 64

Saracens, 144

Scotland, 79–80, 83–84

secret codes, 68

secret documents, 88

   guarded, 101

Secret History, 147

security, 10, 128, 138

shadows, 33, 40, 43, 115

siege engines, 17, 22

sieges, 10, 16, 22, 31, 61

Sinclairs, 79–80, 83

stone, polished, 39

stone galleries, 26

stone monuments, 72

stone tablet, 71

surveys, underwater, 63

survival, cultural, 63

swords, 13, 20–21, 24, 144

symbolic identity, 89

symbolism, coded, 66

symbols
  cryptic, 76

  mystical, 54

  undeciphered, 71

T

Talismans, 39

Templar defense network, 27

Templar emblems, 52

traditions

  cultural, 84

  merchant, 15

  noble, 15

  religious, 7

transcendence, spiritual, 42

transformation, 42, 52, 68, 86

treasure caches, 51

treasure map, 66, 75

U

underwater vaults, 120

unearth Templar relics, 133

V

vessels, spectral, 122

vibrant mythos, 139

vibrant scholarly debate, 101

W

warfare, medieval, 13

Warrior Monks, 145

Made in the USA
Monee, IL
29 September 2025